THE WORLD IN CONFLICT

War Annual 8

Also by John Laffin

Military

Brassey's Book of Espionage
Brassey's Battles: 3,500 Years of Conflict, Campaigns and Wars from A–Z
War Annual 1
War Annual 2
War Annual 3
War Annual 4
War Annual 5
War Annual 6
War Annual 7
Middle East Journey
Return to Glory
One Man's War
The Walking Wounded
Digger (The Story of the Australian Soldier)
Scotland the Brave (The Story of the Scottish Soldier)
Jackboot (The Story of the German Soldier)
Tommy Atkins (The Story of the English Soldier)
Jack Tar (The Story of the English Seaman)
Swifter than Eagles (Biography of Marshal of the RAF Sir John Salmond)
The Face of War
British Campaign Medals
Codes and Ciphers
Boys in Battle
Women in Battle
Anzacs at War
Links of Leadership (Thirty Centuries of Command)
Surgeons in the Field
Americans in Battle
Letters from the Front 1914–18
The French Foreign Legion
Damn the Dardanelles! (The Agony of Gallipoli)
The Australian Army at War 1899–1974
The Israeli Army in the Middle East Wars 1948–1973
The Arab Armies in the Middle East Wars 1948–1973
Fight for the Falklands!
On the Western Front: Soldiers' Stories 1914–18
The Man the Nazis Couldn't Catch
The War of Desperation: Lebanon 1982–85
Battlefield Archaeology
The Western Front 1916–17: The Price of Honour
The Western Front 1917–18: The Cost of Victory
Greece, Crete and Syria 1941
Secret and Special
Holy War: Islam Fights
World War 1 in Postcards
Soldiers of Scotland (with John Baynes)
British Butchers and Bunglers of World War 1
The Western Front Illustrated
Guide to Australian Battlefields of the Western Front 1916–1918
Digging Up the Diggers' War
Panorama of the Western Front
Western Front Companion
We Will Remember Them: AIF Epitaphs of World War 1
British VCs of World War 2

General

The Hunger to Come (Food and Population Crises)
New Geography 1966–67
New Geography 1968–69
New Geography 1970–71
Anatomy of Captivity (Political Prisoners)
Devil's Goad
Fedayeen (The Arab-Israeli Dilemma)
The Arab Mind
The Israeli Mind
The Dagger of Islam
The PLO Connections
The Arabs as Master Slavers
Know the Middle East
Fontana Dictionary of Africa since 1960 (with John Grace)
Hitler Warned Us
Aussie Guide to Britain

and other titles, including novels

THE WORLD IN CONFLICT

War Annual 8
Contemporary warfare described and analysed

JOHN LAFFIN

BRASSEY'S
London · Washington

Copyright © 1997 John Laffin

All Rights Reserved. No part of this publication may be reproduced, stored in a retrieval system or transmitted in any form or by any means; electronic, electrostatic, magnetic tape, mechanical, photocopying, recording or otherwise, without permission in writing from the publishers.

First English Edition 1997

UK editorial offices: Brassey's, 33 John Street, London WC1N 2AT
UK orders: Marston Book Services, PO Box 269, Abingdon, OX14 4SD

North American orders: Brassey's Inc., PO Box 960, Herndon, VA 20172

John Laffin has asserted his moral right to be identified as the author of this work.

Library of Congress Cataloging in Publication Data
available

British Library Cataloguing in Publication Data
A catalogue record for this book is available from the British Library

ISBN 1 85753 216 3 hardcover

Typeset by York House Typographic Ltd, London
Printed in Great Britain by Redwood Books

For Hazelle, as always

Contents

List of Maps

1

Towards the Millennium with Wars Abounding

REINVENTING NATO

'The King is dead, long live the King.' This expression, traditionally used on the demise of a monarch, would be appropriate for NATO. 'NATO is dead, long live NATO.' The North Atlantic Treaty Organisation, founded at the end of the Second World War, had a long and lusty life but with the end of the Soviet Union as a superpower NATO's first generation had passed. It was then necessary for it to be born again, as NATO Mark 2.

NATO Mk 1 had become an irrelevance, a military alliance without enemies. What single potential enemy existed that could threaten a member of NATO, what combination of enemies? NATO Mark 1 could be wound down and broken up but after such a massive investment in money, years, energy, intellectual and political development, that would be a form of waste never before seen in human history.

NATO needed to be reinvented, to be given a new role. It made sense, then, to enlarge itself into eastern Europe to fill any power vacuum. NATO's primary objective should be to secure stability, so that nations may develop in a politically mature and economically sound way. Without an organised expansion, nations could form new and dangerous alliances, with Germany and Russia perhaps as principal targets.

Opponents of expansion point to the obvious fact that the 16 members of NATO have common standards for everything from the calibre of their weapons to the size of their tyres and petrol filler caps on their vehicles. A Rand Corporation Study in 1996 predicted that NATO expansion could cost up to $13 billion to achieve uniformity of equipment for the new NATO countries. And the US, the study pointed out, would pay by far the largest amount.

Does this really matter? It would bring work to many companies and countless individuals but it would not amount to an 'arms race' because the NATO partners are not in the business of aggressive expansionism. One of the successes of NATO has been to sublimate the desire for imperial expansion of the most aggressive European nation of the last 150 years – Prussia/Germany. Perhaps other nations with an imperial itch could be brought into NATO and shown that collaboration is better than confrontation.

The need for a new NATO cannot be overstressed or overexplained. Throughout the period of the Cold War, NATO maintained enormous military forces to keep Europe and the entire Western democratic world protected against the massive threat of the Warsaw Pact countries led by the Soviet Union. As late as 1990 it seemed that the confrontation would last forever and that NATO would remain the bulwark of protection for the free world. The disintegration of first the Soviet Union and then the Warsaw Pact alliance ended all that. There was now no such entity as the Soviet Union, only Russia, as one of a number of countries which made up the 'Commonwealth of Independent States' (CIS). The other Communist states, Yugoslavia, Poland, Czechoslovakia, Rumania, Hungary and Bulgaria shed their dictators and, following a troubled and sometimes bloody process, they became increasingly democratic.

Where then was the need for NATO to continue as a military pact? Where was the enemy, any enemy? Russia's army was humiliated and in a state of decomposition but the nation still had massive weapon stocks and defence systems. Western statesmen and some of their advisors still saw Russia as a possible threat. To reduce this threat, the clever thing to do, they said, was to expand NATO by inviting Hungary and the Czech Republic to join within two years – by the end of 1999 – and other countries could come in later.

Reaction from powerful figures in Russia was extremely hostile. In September 1996 Alexander Lebed, a possible future president of Russia, announced that an expanded NATO was an attempt by Germany to build a Fourth Reich. Considering the bloody destruction inflicted on Russia by the Third Reich in 1941–44 the fears of Lebed and many others were understandable. Igor Rodionov, the Defence Minister, said that Russia was against NATO expansion until Russian public opinion was convinced that NATO was no threat.

At a meeting in Bergen, Norway, on 26 September 1996, NATO defence ministers invited Russia to participate in certain NATO activities. The package seemed attractive.

- Russia would be closely consulted on any NATO-led force operation in Bosnia.
- Russian liaison officers would be welcomed on a permanent basis at NATO HQ in Brussels.
- No NATO rapid reaction operations would become effective without Russian access to planning.
- At the highest level, Russia would be involved in 'crisis management'.
- Russian defence ministers would take part in meetings with their counterparts from NATO countries.

On his visit to NATO HQ in Brussels in October 1996 Lebed said that Russia would agree to 'joint decision-making, joint implementation and joint respon-

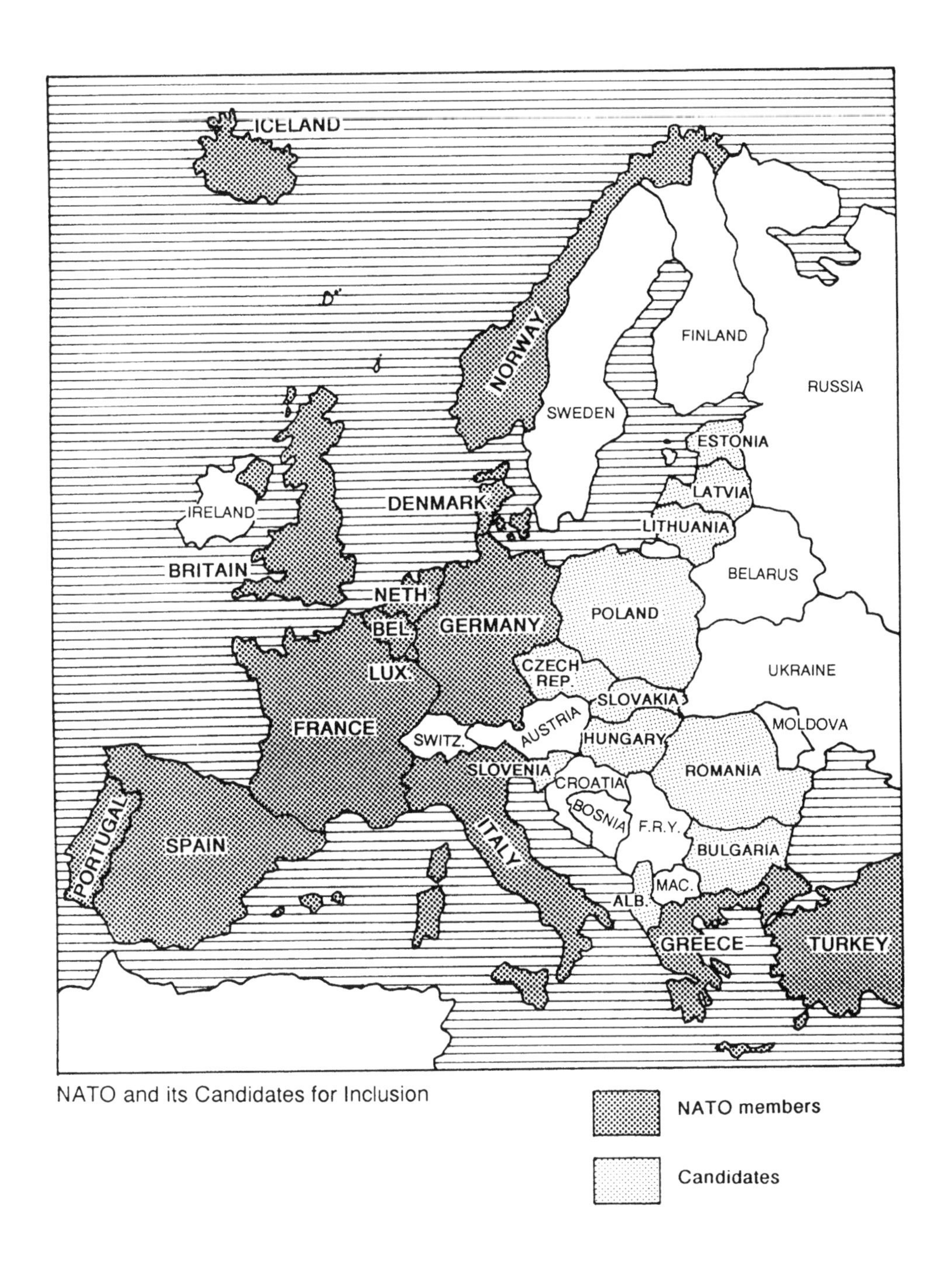

NATO and its Candidates for Inclusion

sibility'. The details would be worked out later. The NATO Secretary-General, General Javier Solana, countered that decisions on enlargement, internal reform and a new relationship with Russia could be ready in time for the NATO summit meeting in mid-1997.

Russia is not the only major country to be worried about an enlarged NATO. Turkey, itself a crucial member of the alliance, is another. Secretary General Solana flew to Ankara to defuse Turkey's threat to veto any enlargement of NATO. The Turkish President, Mr. Demirel, said yet again that his Islamist government would not ratify any NATO agreement to admit new members unless the European Union accepted Turkey's application for membership and delivered aid promised to Turkey.

Trenchant criticism against NATO expansion came from George Kennan, a former US ambassador to Moscow and regarded as a leading expert on superpower relations. NATO expansion, he claimed, would encourage anti-Western sentiment in Russia and lead to a democratic reversal there. Expanding NATO would be the most fateful error of American policy in the entire post-Cold War era.

> It would inflame nationalists and drive Russian foreign policy in directions entirely adverse to Western wishes. Russians are little impressed with American assurances that expansion reflects no hostile intentions. They would see their prestige (always uppermost in the Russian mind) and their security interest as adversely affected. They would continue to see it as a rebuff and would look elsewhere for guarantees of a secure and hopeful future for themselves.[1]

At the time of Kennan's expressions of foreboding, President Yeltsin's chief security advisor, Ivan Rybkin, said publicly that Russia should be prepared to use nuclear weapons if it were faced with a conventional attack. Rybkin said:

> Who will protect us against hotheads abroad? There cannot be any such guarantees. So everyone should be aware that in the event of a direct challenge to our national security we would respond with all available means, with the options including nuclear arms.

NATO had always wanted the right to resort to a limited nuclear strike in the event of being overwhelmed by conventional forces; it was a key element of Alliance nuclear deterrence throughout the Cold War. Now Rybkin was espousing much the same doctrine for Russia.

But could Russia ever again be a real threat? According to the Defence Minister, Igor Rodionov, speaking in Moscow early in February 1997, a lack of funds had damaged morale in the armed forces, reduced combat readiness and badly affected the armed forces' command and control centres. They were in a state of 'horrifying decay' and the reliability of the nuclear weapons system was in question. Rodionov, who left the army to become Russia's first civilian defence minister, said:

> Reform in the armed forces has turned into a struggle for their survival, a struggle against disintegration. Even as defence minister, I am becoming a spectator of the destructive process in the army and can do nothing about them.

He complained that a shortage of satellites meant that Russia's ability to track the nuclear forces of a potential army was deteriorating while the US observed Russia 24 hours a day. The Rodionov assessment of the Russian military decay came as the Clinton administration said that US Senate approval of an international chemical arms ban could help persuade Russia to stop making the deadly nerve gas, A-232.[2]

The date of 27 May 1997 was significant in NATO history for two reasons. First, the Cold War officially ended when the Western Alliance and Russia agreed to co-operate in redrawing the political power map of Europe. President Yeltsin, the NATO Secretary General, Javier Solana and 16 Western leaders signed a co-operation agreement that should clear the way for the expansion of NATO into former Iron Curtain countries. The agreement laid down a mechanism for consultation and co-operation between the Russian government and the Alliance. In addition, it would allow Russia to hasten its own economic reform with the aid of the West.

In effect the agreement, signed in Paris, guarantees any future Russian leader and his government information about Alliance affairs but it withholds any right of veto. Under this agreement, which has the official title of the Founding Act on Mutual Relations, Co-operation and Security between NATO and the Russian Federation, Russia would appoint an ambassador to NATO. Equally significant, Russian liaison officers would have a permanent post in some NATO command headquarters, an extraordinary privilege. The *quid pro quo* is that NATO would have liasion officers in Moscow and other central command centres, though not in regional HQs, such as the giant nuclear submarine base at Murmansk on the Kola Peninsula.

The NATO Secretary General went on record as saying that the new agreement with Russia would not prevent NATO from developing military infrastructures in the new member countries, regardless of Russia's views. It is important that in the agreement NATO states that 'we have no intention of deploying nuclear weapons or substantial conventional forces on the territory of the new member states'. However, 'substantial' could mean anything, as can the phrase 'no intention'.

The second major event that made 27 May 1997 so significant was a statement by President Yeltsin. Having requested a five-minute unscheduled break in the ceremonies after the signing of the Founding Act, Yeltsin said, 'I have today taken the decision to dismantle all nuclear warheads from missiles directed at countries represented around the table'. This was an astounding concession and one which, apparently, took the world's Intelligence services by surprise.

Yeltsin and his ministers had hoped for certain concessions in return but they were not forthcoming. For instance, they wanted a legally binding treaty status and an absolute ban on NATO nuclear arms on its borders. The parties to the Act remain divided on the potential for NATO to expand to include the Baltic states of Latvia, Estonia and Lithuania.

By 1999–2000 we might well have a NATO Mark 3.

THE SIPRI REPORT: TOO OPTIMISTIC?

The Stockholm International Peace Research Institute (SIPRI) produces an annual report, whose authors generally try to appear relieved in their analysis of what has happened in the previous year and optimistic about the year ahead. The SIPRI report published in mid-1996 may have been a little too relieved and too optimistic.

The report spoke of the 'striking success' of UN, US and European efforts in ending the conflict in Bosnia-Herzegovina, as a result of the signing of the Dayton Agreement. SIPRI also drew attention to the end of the conflict between the Croation government and the Croation Serbs. Again, there was an agreement following military victories.

Unfortunately each military victory also has a corresponding military defeat and throughout history defeated nations, communities, tribes and groups have sooner or later sought revenge and 'justice'. That a strong foreign military presence is necessary in the dismembered Yugoslavia is evidence that some conflicts have not ended; they are merely in suspension. As just one instance, in January 1997 conflict broke out after 26 Muslim families were expelled from the Croat-controlled section of the town of Mostar.

At the same time, large-scale troop movements by both Muslim and Serb units led to dangers of war over the disputed town of Brcko. The NATO-led Stabilisation Force ordered both sides back to their barracks. Brcko, on Bosnia's northern border with Croatia, is strategically important to both sides. Neither Muslims nor Serbs would agree to the fate of the town during the negotiations for the Dayton peace agreement, so a decision was left to arbitration proceedings.

In Mostar, Edward Joseph, head of the regional centre of the Organisation for Security and Co-operation in Europe (OSCE), said that Muslim-Croat hostilities over evictions of Muslims was an attempt by extremists to destroy the peace process in Mostar.

The senior Western mediator in Bosnia, Carl Bildt, said: 'Nobody can gain anything from a new war for control of Brcko. I am not sure there were any winners in the past war and I am sure there would be no winners, only massive losers'. But common sense urgings mean absolutely nothing to embittered people and anger over Brcko could well precipitate a war by 1998.

Muslims, Croats and Serbs were all talking about ethnic cleansing, another indication that people of the old Yugoslavia live with an expectation of conflicts to come. To declare that any war is over because no major campaign or battle

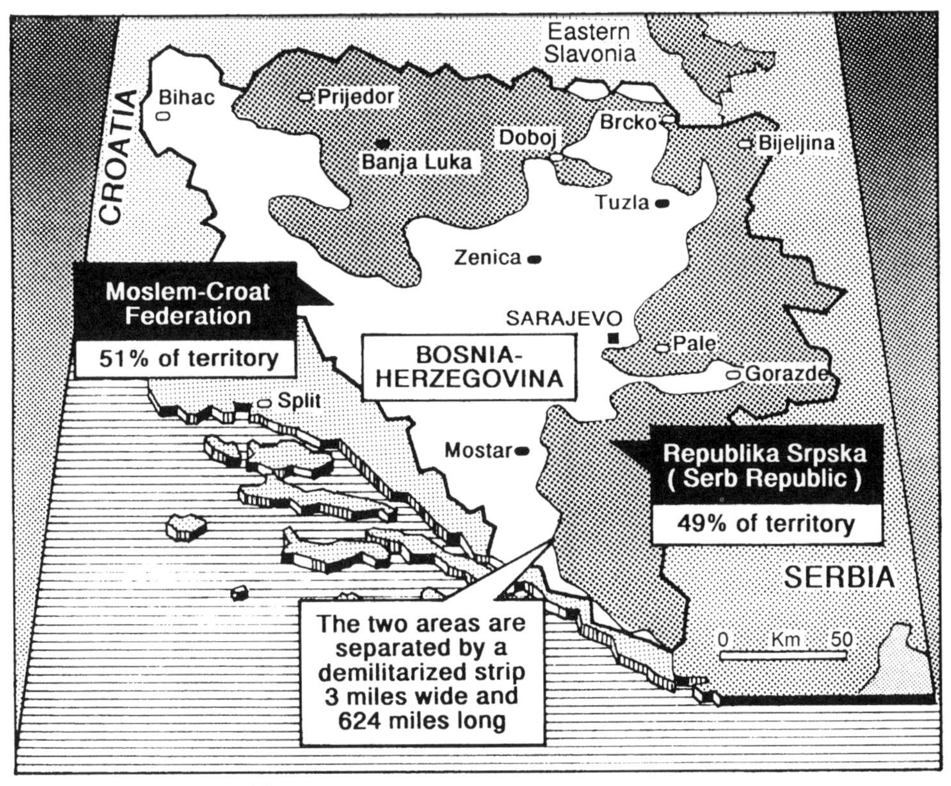

The Dayton Division of Bosnia

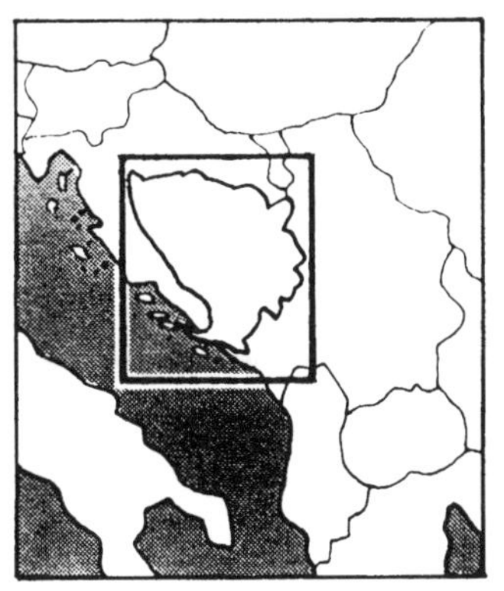

has taken place is unrealistic. The entire area of Serbia, Bosnia-Herzegovina, Croatia, Montenegro and Kosovo will remain a powder keg for years.

Other parts of the Balkans are flash points for civil wars – Albania and Bulgaria in particular. This is a crisis zone for the West, as, in the Aegean Sea, Turkey and Greece are making reciprocal threats of war and gestures of war and sounding as if they mean them.

The SIPRI report expressed concern that the prospects for a new NATO peace mission were unclear because of the organisation's ever-deepening financial crisis. The most effective action in peacemaking, according to SIPRI, was that taken by the greatest political and military power, the US, and to some extent by Russia.

Since publication of the report, the US has agreed to pay the billions of dollars it owes the UN, a step that was apparently linked to the end of Boutros Boutros-Ghali's tenure as UN Secretary General.

The SIPRI researchers drew attention to the phenomenon that all the major armed conflicts were internal rather than between states.[3] This was so, even though outside forces were involved in some of the wars. For instance, in Tajikistan, forces from the CIS – mainly Russian troops – were used against the opposition. And in Liberia, peacemaking corps took a major part. This is described elsewhere in this book.

SIPRI research showed that aggregate world military spending continued to decline. Indeed, this has happened every year since the end of the Cold War. But the simple observation that military spending was in decline is unrealistically comforting because spending in the Middle East, North Africa, South-East Asia, India and Pakistan – all unstable areas – has increased. The SIPRI report's most alarming statistic concerns arms exports – $22.8 billion for 1995.

Sooner or later governments and armies, afflicted with massive arms indigestion, become bellicose and then belch into aggression. Syria and Iran were among the countries who were in this condition by the beginning of 1997.

TERRORISM: 'THE WAY OF WAGING WAR'

In October 1996, the Chairman of the US Joint Chiefs of Staff (JCS), General John Shalikashvili, made one of the most significant public statements by a JCS chief for several years. Speaking to more than 1,000 senior military and civilian officials at the Pentagon, Shalikashvili said: 'Tactics have changed so much that terrorism is no longer an adjunct to something else. In many cases, terrorism will become the way of making war against us.'

The General probably meant strategy as much as he meant tactics and what he went on to expound was acutely important. He obviously thought so, since he had actually called the meeting; he was not making an entertaining address to a Rotarians' luncheon. According to Shalikashvili, anti-terrorism, Joint Vision 2010 (a US defence programme) and the Quadrennial Defence Review

(QDR) were the main priorities for US defence officials. For a long time, he said, the US had depended on Britain and Israel to provide the world's best experts on anti-terrorism, in both intelligence and operational terms. They were still as good as their reputation but the time had come for the US Department of Defense to produce its own experts, men and women who would be so efficient that Israel would seek expertise from the US, as much as the other way around.

Why was the US facing a war of terrorism? Shalikashvili explained that the US was so comprehensively efficient in conventional warfare and its power so overwhelming that 'adversarial nations' could penetrate American defences only through terrorism. He did not specify the adversarial nations but Pentagon briefing papers include many references to Iran, Libya, Syria, Iraq and enemies such as Islamic fundamentalists.

The Shalikashvili lecture and the vigorous, almost impassioned way in which the general delivered it startled his listeners and the 'friendly' military attachés who were privileged to hear it. One of them told me: 'I came away with the uncomfortable conviction that Americans are terrified about foreign terrorism and that they are aware, at last, that they know very little about dealing with it'.

Despite this criticism the US has several counter-terrorism units, including specialists for dealing with chemical and biological (CB) alarms. They are:

- Technical Escort Unit (TEU): international, instant reaction for intelligence, recovery, decontamination and make-safe operations
- Chemical/Biological Anti-Terrorism Team (CBATT): responsible for development of equipment and technology
- Army Radiological Advisory Medical Team and Radiological Control
- Army 52nd Ordnance Group: quick reaction against unconventional terrorist weapons
- Navy/Marine CB Incident Response Unit
- Air Force Radiation Assessment Team
- Air Force Technical Application Centre: instant reaction to CB incidents
- Armed Forces Radiobiology Research Institute: a medical unit dealing with radiation exposure.

Terrorism as 'the way of waging war' was demonstrated in Israel and Algeria in August 1997. *Hamas* suicide bombers exploded two bombs in a Jerusalem market place, killing 30 people and wounding about 180 others. Some Palestinian spokesmen said, in as many words, that as the Palestinians could not hope to win a 'normal' war against Israel, terrorism was their only viable alternative.

In Algeria, torn apart by a civil war, terrorism by the State and by the forces of Islamic fundamentalism is certainly the way of waging war, even if not quite in the way that General Shalikashvili meant.

THE CIA REPORT ON CHEMICAL WEAPONS

The US Senate Intelligence Committee early in 1996 asked the CIA for a report on Iran's non-conventional weapons programme. The Agency's analysis was submitted in November and its main statement was alarming.

> Iran's chemical warfare programme is already among the largest in the world, yet it continues to expand. Also, it is increasingly diversified, even following Iran's signing of the Chemical Warfare Convention in January 1993.

The Iranian biological weapons programme involves both toxins and live organisms; furthermore it has the technical infrastructure to support a biological warfare programme and needs little foreign assistance. This is disturbing because it means that the US and its allies have no way of putting pressure on third countries who might supply Iran and their opportunities for intelligence-gathering are fewer.

Other statements from the report are:

- Because of the dual-use nature of biomedical technology, Iran's ability to produce a number of both human and veterinary vaccines gives it the capability for large-scale BW agent production.
- Iran has a stock pile of several thousand tons of CW agents, including sulphur, mustard, phosgene and cyanide gases. The country can produce a further 1,000 tons each year.
- Iran is developing a production capability for the more toxic nerve agents and it has the means of disseminating these agents through artillery, mortars, rockets, aerial bombs and Scud warheads.

In a supplementary, restricted report, the CIA told the Senate Committee that the threat from BW and CW weapons was 'much more immediate' than that from nuclear weapons. The Agency analysts said that while Iran was vigorously pursuing research and development it would need considerable foreign assistance to build a viable nuclear weapon before the end of the century and perhaps not even by then. However, the report concluded, influential ayatollahs and military leaders in Iran considered that they could inflict greater damage against their perceived enemies by means of CW and BW than through nuclear devices.[4]

General Shalikashvili, like his political masters, has identified North Korea and Taiwan as Asian flashpoints. The region is of so much concern to the US that at the beginning of 1997 the Americans had 100,000 troops in East Asia, about a third of them stationed on the Japanese island of Okinawa. In the Sea of Japan or close to it is the US Seventh Fleet, built around USS *Independence*, the massive aircraft carrier with its crew of 5,000. The ship's captain told a British journal 'I have on this ship 5,000 of the toughest Americans you're going to find anywhere ... These guys eat nails for breakfast'.[5]

Since the end of the Cold War it has not been quite clear why the US maintains such great military strength in this region. The Soviet Union, now dismembered, is no longer the threat that it was. Perhaps the Americans magnified the threat in the first place. The Clinton administration might not like to say so in as many words but China is now the threat – not directly against the US, but possibly against Japan, Taiwan and South Korea. The presence of the American armada and army, together with their air force support, is a stabilising influence – at least, this is how American war-thinking goes. It is just possible, though, that the American military presence could become an intolerable provocation to China, whose strength is on the rise. There is another problem: Japan is militarily weak and because it is defended by the US its politicians have no great incentive to make their country strong. The other side to this coin is that it may be safer for the world if Japan were to stay weak; the rapacious and brutal empire-building which Japan engaged in during the Second World War is still a major feature in international memory.

The death of China's leader Deng Xiaoping on 21 February 1996 ensured that the Clinton administration and the Chiefs of Staff would not wish to weaken the American stance in the Asian sphere, at least not until the intentions of Deng's successor, Jiang Zemin, and his senior political associates become clear.

THE BLUE BERETS: SHORT OF CASH, VICTIMS OF THEFT[6]

During 1996, 81 countries provided troops for the UN but the largest contingents by far came from Pakistan (1,704 troops), Russia (1,215), India (1,028), Bangladesh (1,178), Jordan (1,127), Brazil (1,113), Poland (1,065), Canada (1,060), Uruguay (911), and Finland (910). (In peacekeeping generally Britain and the US provided more troops than other countries but not as UN Blue Berets.)

In 1994 the UN carried out 22 peacekeeping missions but in 1996 the number had declined to 16. There was a distinct co-relation between the number of troops and the amount of money available. The UN's Department of Peacekeeping Operations (DPKO) had only $1.3 billion to spend in 1996, compared with $3.5 billion in 1994.

Having failed abjectly in Somalia, Rwanda and Yugoslavia-that-was, the UN Security Council was cautious about taking on new commitments, although as soon as the UN Secretary General, Kofi Annan, came to office on 1 January 1997 he announced that peacekeeping would remain a major priority.

The UN remains the most credible peacemaker but it is not the only one. The US, Britain, France, the Netherlands, Italy, Denmark and Spain all contributed troops to the policing operations in Bosnia-Herzegovina but under NATO command.

The UN operations during 1996-97 were:

Angola: 7,138 personnel.

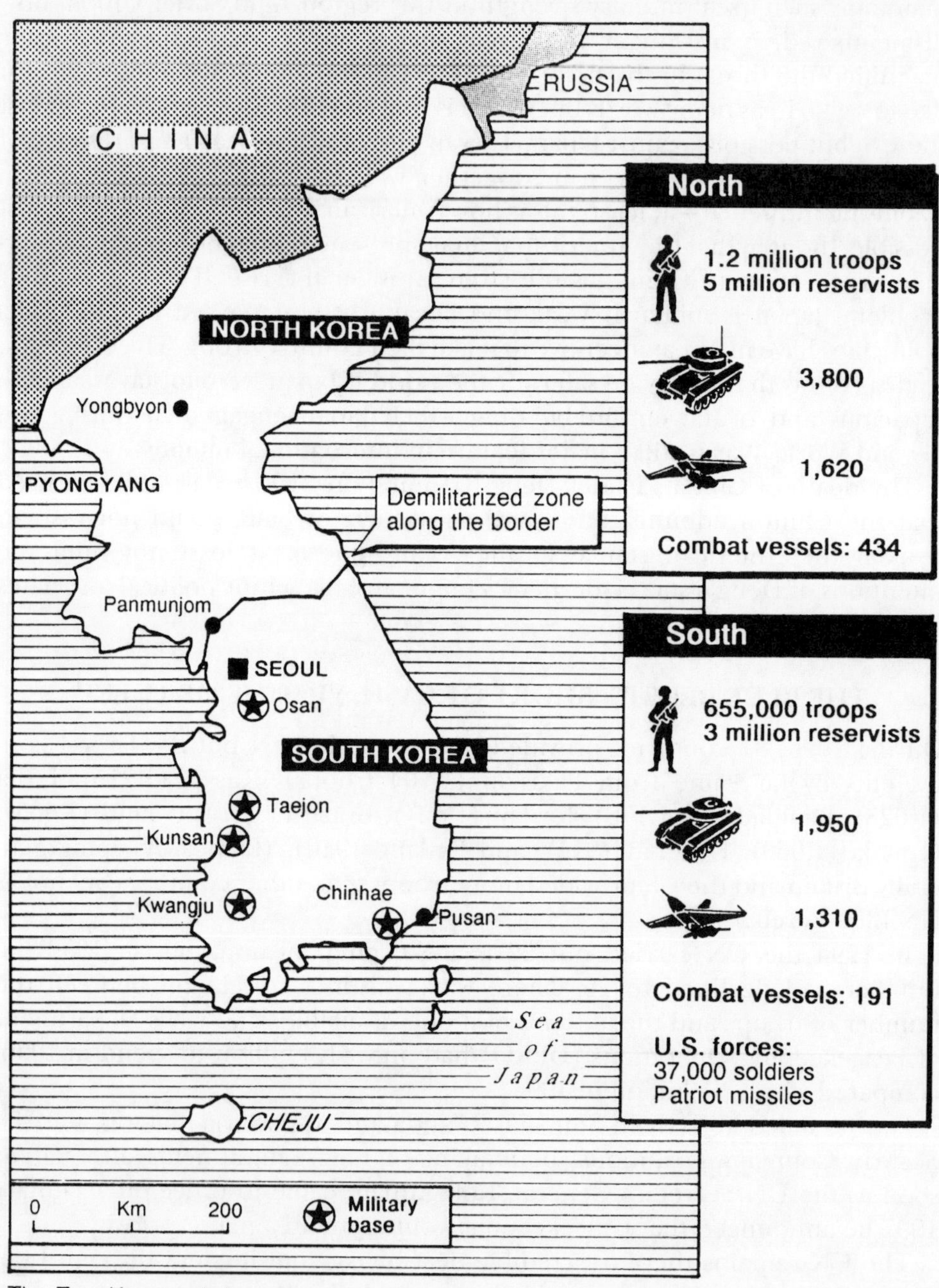

The Two Koreas: A face-off
(South Korean Statistics)

Bosnia-Herzegovina:	1,707 personnel.
Croatia:	UN Missions of Observers in Prevlaka: 27 officers.
Croatia:	UN Transitional Administration in Eastern Slavonia, Baranja and Western Sirmium: 5,405 personnel.
Cyprus:	UNIFICYP: 1,197 personnel. In place continuously since 1964.
Georgia:	Observer mission: 124 personnel.
Golan Heights:	UN Disengagement and Observer Force: 1,053 personnel. In position since 1974.
Haiti:	UN Support Mission: 800 personnel. Separately, the US has other forces in place.
Iraq/Kuwait:	UN Iraq-Kuwait Observation Mission (UNIKOM): 1,112 personnel.
India/Pakistan:	UN Military Observer Group in India and Pakistan (UNMOGIP): 50 personnel. Established in 1949 and continuously on duty since then.
Lebanon:	UN Interim Force in Lebanon: (UNIFIL): 4,526 personnel.
Liberia:	UN Observer Mission: 70 personnel.
Macedonia:	UN Preventive Deployment Force: 1,097 personnel.
Middle East:	UN Truce Supervision Organisation (UNTSO) 163 personnel. Established in June 1948. A low-key but continually operational detachment of about 150.
Tajikistan:	UN Mission of Observers in Tajikistan: 44 personnel.
Western Sahara:	UN Mission for the Referendum in Western Sahara: 44 personnel. Established in 1991, MINURSO was expected to conclude its role by 1993. It was still there in 1997.

One of the greatest hazards faced by UN peacekeepers is the theft of their equipment and funds. Probably the largest robbery took place in Somalia in 1994 when $3.9 million was stolen and never recovered. UN security officers in New York have strong suspicions about the identity of the culprits but since high-ranking officers are thought to be involved no names have been mentioned in public. When the UN brought its mission in Yugoslavia to an end in 1996 millions of dollars worth of equipment 'disappeared' – vehicles and tanks, electronic systems, computers, sophisticated radios and television surveillance systems and furniture. Everything that the UN has taken into a mission is supposed to leave with it but virtually nothing does. Sometimes essential equipment is stolen *during* a mission, so that peacekeepers' lives are put at risk.

SHIRBRIG: MODIFIED RAPTURE

The often-voiced criticism that the UN lacks a quick-response reaction-force that could be speedily moved to a trouble spot was taken seriously by seven countries which, late in December, formed the Stand-by Forces High-Readiness Brigade (SHIRBRIG).

Meeting in Denmark on 15 December 1997, the defence ministers of Canada, Poland, Denmark, Sweden and Norway, as well as the Chiefs of Staff of Austria and Holland, signed an agreement to create the brigade of 4,000 troops. While it is far from perfect, SHIRBRIG was praised by the Canadian defence minister as 'the most promising programme available to provide the United Nations with a rapid reaction capability at the tactical level'.

SHIRBRIG was given an HQ staff of just 17, based at Hoevette Barracks, north of Copenhagen, and its first commander was Brigadier General Finn Saermark-Thomsen of Denmark. The brigade, which will be fully operational by 1 January 1999, will be in unified formation only for operations and training periods.

SHIRBRIG's shortcomings are obvious. It will need a month to deploy and will be used only for crises where there is little danger of fighting breaking out. Defenders of the concept say that the very presence of the brigade could be enough to keep the peace. There is a further built-in problem: each country is permitted to opt out of a particular operation, though without preventing the rest of the brigade from taking part.

However, withdrawal by one or more countries would reduce the strength of the brigade and might well lessen the resolve of the others to act in a positive way. The forces allocated – proportions had not been worked out in early 1997 – will stay in their home countries until the UN Security Council authorises a mission, which must then be approved by the respective governments. My belief is that SHIRBRIG will be hamstrung by its rules of engagement, but it could provide a specimen on which later ready-forces could be modelled.

References

1. Kennan was writing in *The New York Times,* 6 February 1997.
2. Rodionov was speaking to a Russian media conference. His frankness was remarkable and unusual. To an extent, he may have been trying to distance himself from blame for the sorry state of the Russian armed forces.
3. Peter Steele, a spokesman for the UN High Commission for Refugees, said in December 1996, that a UN 'disasters report' of 1994 had 'highlighted the fact that of the 82 major conflicts that had taken place during the previous decade 79 were within states rather than between states'. On this point, UN and SIPRI research agree but sometimes dispute within a state can involve outside powers to a dangerous degree. The wars within Yugoslavia-that-was are a significant example.
4. In August 1996 Barbara Starr, US correspondent for *Jane's Defence Weekly,* revealed that the Americans secret service and the army were collaborating on ways of protecting the American president from CB terrorism.
5. Captain Fellin was speaking to Richard Lloyd Parry of *The Independent,* London. The officer's bravado, while reflecting pride in his crew sends unfortunate signals when translated into the

languages of countries which fear 'American imperialism'. Also, such language is used by propagandists trying to stir up anti-US sentiment.

6. The statistics in this section come from a report issued by the UN's Department of Peacekeeping Operations (DPKO) at the end of 1996. The numbers of personnel involved in various missions can vary from month to month. An unhappier statistic is that at the end of January 1997, 42 peacekeepers had been killed and 315 wounded by anti-personnel landmines, mostly in Yugoslavia-that-was. There are landmines in 68 countries, an increase of four on the previous year. The UN has mine-cleaning programmes and has hired engineers to clear about 100,000 landmines a year but every year more than two million are being laid. According to Peter Hansen, Under-Secretary General for Humanitarian Affairs, it will take about 1,000 years to clear all the mines buried in Afghanistan alone.

2

Afghanistan and the Taliban

Afghanistan has always been an unruly, fractious and disunited country, plagued by enmities. Its various clans had nothing in common except a hatred of foreigners. Militarily, it was a nightmare for tribal chieftains waging war and even more so for invaders. Its great mountain ranges, harsh summer and freezing winter climate and tortuous communications made properly-organised operations difficult and often impossible. Yet this was the country which the Soviet Union unwisely invaded in December 1979.

The Russians, centuries before there was a political entity called the Soviet Union, had dreamt of gaining a warm-water port on the Indian Ocean but this could only be achieved by conquest of Afghanistan. The difficulties of conquering and then controlling Afghanistan had been demonstrated by the British, who fought three wars there – in 1842, 1879–80 and 1919. In the end the British, those generally indefatigable imperialists, gave up.

The leaders of the Soviet Union, intent on creating a great Communist empire, continued with their expansionist policy by making Afghanistan a satellite. The 1979 invasion was presented as a friendly move 'to help the Afghan government maintain control over rebellious elements'. The government was a puppet one, led by President Hafisullah Amin, who was killed early in the fighting. The Soviet government replaced him with Babrak Karmal but he was too irresolute for Moscow and Dr Muhammad Najibullah, head of the KHAD, the ruthless secret police, was given the presidency.

The war, then, was fought by the combined Soviet and Afghanistan armies on one side and the Mujahideen Resistance, inspired by Islamic fervour, on the other. Weapons poured into Afghanistan, the Soviet Union supplying the army, while the United States and Pakistan aided the Mujahideen. Later the British also gave weapons to the Mujahideen, which consisted of at least 20 different groups. The main ones were:

- *Jamiat-i-Islami,* led by Ahmad Shah Massoud
- *Hezb-i-Islamic,* led by Younis Khalis
- *Hezb-i-Islamic* Independents, a breakaway faction under Gulbuddin Hekmatyar

- *Jawzjani* Uzbek Army, led by General Abdul Rashid Dostam
- *Ittehad-i-Islami,* whose leader, Muhammad Sfzal, was backed by the Saudi government
- *Karakat-i-Inqilib,* a group commanded by Nabi Muhammad
- National Islamic Front of Afghanistan, a strongly religious body whose leader was Pir Sayyed Ahmed Gailani
- National Front for the Rescue of Pakistan, created by Sigbatullah Mojaddedi, who favoured the return of the monarchy to Pakistan

Officially, the Communist party was known as the People's Democratic Party of Afghanistan (PDPA) but the term 'democratic' was meaningless by western values. Comrade Najib – a label favoured by President Najibullah – led the PDPA, but it was an amalgam of the Parcham and Khalq factions and was therefore unstable. Indeed, the two factions detested each other.

In the Soviet Union the military campaign in Afghanistan was presented as a victory but in truth it was a defeat. Even more, it was a political embarrassment. Public opinion, long ignored in the Soviet Union, had a decisive effect in bringing about the Soviet withdrawal, which was ordered by President Mikhail Gorbachev. It came into full effect in April 1989.

The renewed conflict was now between Najibullah's army of the Democratic Republic of Afghanistan (DRA) and the Mujahideen Resistance. The army, with its vast Soviet stockpile of aircraft, tanks and weapons, was powerful already. Now Najibullah enlarged the paramilitary KHAD to 35,000 and created a special guard of 11,000 to protect his person. Even so, his own Minister of Defence, Nawaz Tanai, acting for his own 'People's Democratic Party of Afghanistan', attempted to overthrow him.

Meanwhile, the Mujahideen leadership, heeding its inept Pakistani advisors, had turned from guerrilla tactics to conventional warfare. This led to a major defeat when they besieged Jalalabad in 1989. Based in the Pakistani city of Peshawar, the Resistance groups plotted against one another in an amazing display of self-destruction. At least 34 leaders were murdered.

In October 1991 the Mujahideen, with several hundred captured Iraqi tanks given to them by Saudi Arabia, attacked Gardez and in May the following year they routed the defenders of Kabul. This success was the result of the strength which came from the alliance of three major Resistance leaders – Ahmad Shah Massoud of Jamiat-i-Islami, Sayyed Mansour Nadari, overlord of Baghdan Province and General Abdul Rashid Dostam, commanding the Uzbek militia. They were supported by General Abdul Mohmin, whose 70th Brigade of the Afghan army had mutinied against Najibullah's government.

Almost at once in-fighting took place, with Massoud pushing the supporters of Gulbuddin Hekmatyar out of Kabul. The only one of 51 leaders weak enough not to be considered a threat to the others was Sigbatullah Mojaddedi, who was appointed President, with Massoud as Defence Minister. President Buranuddin Rabbani succeeded Mojaddedi.

Najibullah and his brother found sanctuary of a sort in the UN compound in Kabul. All his men had either deserted to the Mujahideen or had fled as refugees and 'Comrade Najib' was no longer considered a threat. That the Mujahideen leaders allowed him to go on living had nothing whatever to do with their humanity; to them he was now totally insignificant.

The Taliban Farsi – New Model Army

In War Annual 7 I described the foundation in 1994 of the *Taliban Farsi* – meaning 'Seekers of the Truth' – by retired officers from Pakistan's Inter-Services Intelligence (ISI). These officers had earlier backed Hekmatyar but they had painfully and embarrassingly learnt that his intransigence had caused much of the endless violence. With the connivance of the Pakistan government, the ex-officers secretly recruited men for Taliban among the million or more Afghan refugees in the Pakistan province of Baluchistan and the North-West Frontier. The Western media has generally presented the Taliban's fighters as 'students' but this label is true only in that all Muslims are students of the Koran, and many Taliban fighters had been attending *madrasahs* or religious schools in Pakistan and in the remote south-west of Afghanistan. The Taliban's soldiers were not and are not students in the Western meaning of they word, though no doubt many of these men might have attended ordinary universities had not the war destroyed every aspect of normal Afghani life.

American, British and Russian intelligence services did not know of the Taliban in 1994 though the Iranian government did and sought to influence its politics by infiltrating its ranks with mullahs sent from Teheran. They probably had little success in Taliban's command structure. The leader, a cleric named Maulana Muhammad Omar Akund, and his senior supporters were as zealous and fundamentalist as any Iranian mullah.

From their southern base, the Talibani quickly swept through nine provinces and in February 1995 the movement's vanguard stormed Hekmatyar's HQ at Charasiab, a mere 15 miles south of Kabul. Remarkably, Hekmatyar's garrison did not fire a shot and, together with 6,000 Uzbek militiamen who would normally have fought ferociously, withdrew to the mountains. The failure of the enemy to stand and fight the Taliban was the result of an unusual propaganda campaign carried out by unarmed 'holy men' advancing ahead of the Taliban's fighters. These 'mullahs' – for lack of a better word to describe these emissaries – warned Hekmatyar's men and the Uzbeks that to fire on the Taliban would be an outrage against Allah. Passages from the Koran were quoted to support this assertion. 'Shooting a Taliban soldier is like a Catholic shooting a priest', a Western aid worker told a journalist.

Since the outside world knew little about the Taliban, the UN Security Council sent a special envoy, a former Tunisian foreign minister, Mahmoud Mestiri, to find out more about the Taliban's leadership, arrange a formal ceasefire and follow this up with a 'peace process'. Mestiri's mission was abortive. While President Rabbani was willing to meet Mullah Omar of the

Taliban it was clear that Omar would not meet him. 'Peace is not part of our agenda', Omar told Mestiri. 'The guilty men must be slain.'

Indeed the Talibanis were carrying out numerous executions. Several thousand men with 'bad records' may have been killed in 1995–96. At the same time, the Taliban was establishing *shuras* or ruling councils to administer the law in cities and towns. When an example was needed to impress an unwilling populace the Talibani hanged a few men of rank, tied their corpses to the raised barrels of tank guns and drove around the streets.

Reporting to the UN Secretary General, Boutros Boutros-Ghali, Mahmoud Mestiri said: 'The Taliban's leaders know exactly what they want and are absolutely determined to get it'. Mestiri went on, 'The Taliban is not interested in conquest in itself but in social and economic justice'.

The ordinary people of Afghanistan were desperate for such reforms, but it became all too obvious that the Taliban's ideas of social and economic justice were not exactly identical with those of a cultured, educated and Westernised Tunisian, Mestiri.

The War in 1996–97

At the very end of 1995 President Rabbani began a 'peace offensive' by offering to visit all the main Afghan rebel groups in their strongholds, a conciliatory and even courageous gesture. While he did not personally travel to see General Rashid Dostam of the *Jumbish-i-Milli* party, he sent a delegation to explore peace. He even offered negotiations with the Taliban and said that he would visit the leadership in Kandahar. Mullah Omar's initial reaction was to reject negotiations out of hand: 'they are a smoke-screen for military operations', he said.

Nevertheless, he softened this stance, perhaps because of overtures from the deputy foreign minister of Iran, Alahuddin Brojerdi, who went to Kandahar. Various signs of moderation reflected the military impasse and the huge costs of the conflict. The Red Cross warned, early in 1996, that the blockade of Kabul was reducing food and fuel to the point where 'one of the world's great humanitarian disasters' was imminent. The Taliban's bombing of Kabul was also putting pressure on Rabbani's administration. In December–January about 100 civilians had been killed and 400 wounded, according to the Red Cross.

Despite repeated attacks, neither Dostam's Uzbeks nor the Taliban made any great inroads into Kabul's defences. This was because of Ahmah Shah Massoud's superior generalship. Going on the offensive against Dostam, Massoud units captured a strategic mountain pass near the Salang highway. Also, Massoud used his air force effectively against the Taliban, which no longer looked invincible. They had about 50 aircraft of their own but they had no more than 10 trained pilots and not until early 1996 did they set about a systematic training programme for tank crews.

A tremendous explosion near Kabul destroyed at least 50 truckloads of arms

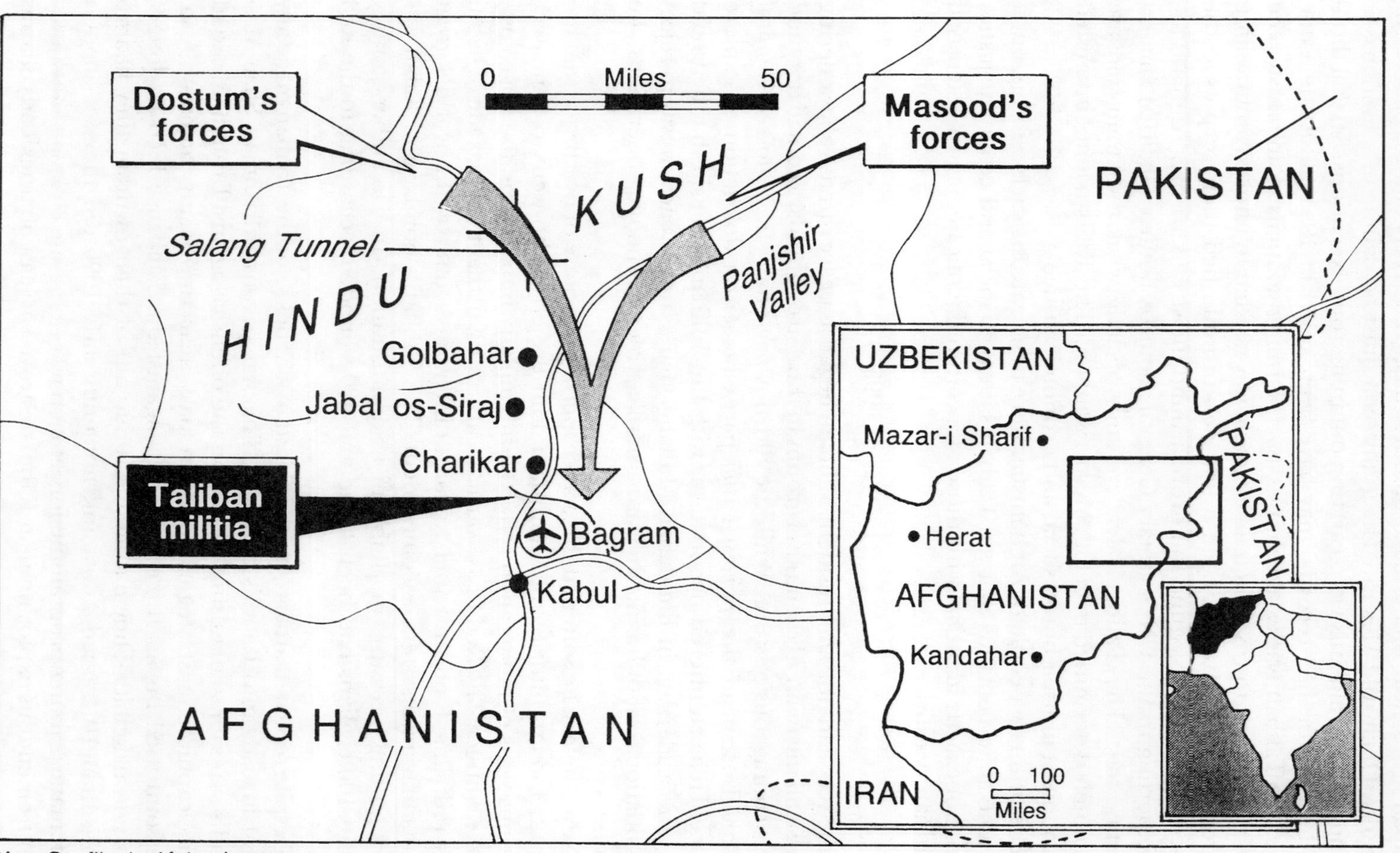

New Conflict in Afghanistan

and killed several Taliban leaders. The explosion was attributed to 'sabotage' but it could equally well have been caused by inept handling of munitions.

Rabbani was militarily inexperienced but Massoud knew very well that if the Taliban and Dostam co-ordinated their efforts and attacked on all fronts, the government and its army could not hold Kabul. Meanwhile, Rabbani worked assiduously at securing some form of power-sharing agreement. A Western ambassador told me, in February 1996, that the probable outcome would be that the government would hold Kabul, and the west and north of Afghanistan, while the Taliban would control the south and east. He admitted that Hekmatyar and Dostam could well overthrow the government in one of their frequent coup attempts.

The basic problem in early 1996 – as in earlier decades – was ethnic discord. The Tajiks, represented by Rabbani, the Uzbeks led by Dostam, and the Pathans, who formed the great bulk of the Taliban, all demanded that they should form the majority in an interim government. On this point the Taliban was the most insistent. Any success achieved by one element had the immediate effect of making the victor more militant. Rabbani was often said by the Western media to be the most ready to compromise; after all, had he not offered to surrender power to a UN-sponsored interim government? Indeed he had, but when Massoud delivered him a battlefield victory Rabbani at once revoked his offer of compromise.

Possibly the best chance Afghanistan had of peace would come from the diplomatic pressure exerted by Pakistan, Iran and Uzbekistan, the United States and Russia. Various mediators suggested that the Afghan king, 82-year-old Zahir Shah, be invited to return to his country as the only symbol of national unity whom all Afghans would respect. From Rome, the King himself let it be known that he would return but only if the leaders of all the factions agreed.

In March 1996 some observers predicted that Kabul would fall to the Taliban. A Red Cross official in the capital told me that such an outcome would mean 'the end of civilisation' in Afghanistan. Probably 95 per cent of the populace of Kabul (1.2 million) were cold and starving, he said. When I put this to UN officials in New York I was told that an airlift of food was beyond UN capability; $124 million was necessary to meet the emergency but other countries had offered only $20 million.

By June, with the Taliban push apparently stalled, Massoud, the real power behind Rabbani and the entire administration, was working at forming a coalition. He made the huge concession of inviting Hekmatyar, his great rival, to be Prime Minister. This showed Massoud's political acumen, for Hekmatyar was no longer a military threat; the Taliban had seen to that. By giving him a position of great political power Massoud was also breaking up the alliance between Hekmatyar and General Dostam.

To some extent Massoud's negotiations paid off; in June Hekmatyar assumed the role of Prime Minister. Also, Dostam agreed to open a supply route to

Central Asia, so that Kabul and regions in the north were no longer wholly dependent on the main road from Pakistan.

Early in September the Iranian deputy Foreign Minister, Alauddin Brujerdi, proposed to his Pakistani counterpart, Nazir Sultan, that a regional summit be held to bring an end to the four-year civil war. Others who would attend included foreign ministers from Tajikistan, Turkmenistan and Uzbekistan as well as factional leaders from Afghanistan.

Massoud's plan was complex. First, while Rabbani was still in office, the coalition government would draft a new constitution, create a new national army and set up all-party commissions to get the country on its feet. Following this, Rabbani would voluntarily resign, a ceasefire would come into effect and elections would be held, possibly under UN auspices.[1]

In the midst of Massoud's tortuously patient negotiations the world was treated to an event of high drama. In July 1995 a Russian Ilyushin-76 cargo plane was flying munitions from Albania to the government in Kabul when Taliban jets intercepted it and forced the pilot, Captain Vladimir Sharpatov, to land at Kandahar. He and his crew were held hostage, though not ill-treated. On 18 August 1996, Sharpatov and his co-pilot, Kazinur Khairullin, persuaded their Taliban captors that the crew should service the aircraft if it was to remain serviceable for Taliban use. Under this pretence, they started the engines and took off just over the top of cars that had been rushed up to block the getaway. The Ilyhushin flew at treetop height before reaching the safety of Iranian air space and touched down at Sharjah in the United Arab Emirates. The Russian people were ecstatic about the dramatic escape and President Yeltsin awarded the crew medals for bravery. He knew, as the airmen knew, that the affair had begun as a humiliation for Russia. The daring escape had transformed it into a triumph.[2]

Jalalabad – and on to Kabul

All the moves towards some form of peace assumed that the Taliban would co-operate but at the end of the second week in September Muhammad Omar and his lieutenants showed that they were not impressed. They captured Jalalabad, 70 miles east of Kabul on the main road to Pakistan. By cutting this supply route they controlled two-thirds of the country at a stroke. The Afghan army commanders publicly announced that the Taliban's success was of no great importance. One divisional general boasted to journalists that he would buy them lunch in Jalalabad or shave off his beard in disgrace. He should have known better than to seek such dangerous publicity.

Certainly, the battle for Jalalabad was not a genuine trial of military strength. The city's defences were poor and government troops had arrived too late to offer stiff resistance. However, the city's capture had strategic importance, for the government was now desperate for outside supplies. The one man who could provide them was General Dostam, who controlled the only link with the outside world along the mountainous road to the countries of Central Asia.

Dostam was hardly dependable – he had already changed sides four times.

The Taliban took over the rich smuggling activities of Jalalabad, providing it with much-needed funds. The 'students' seized the thousands of luxury cars stolen from Europe and the Middle East and waiting in guarded compounds for delivery into Pakistan. Jalalabad is the main city of Nangarhar province, which grows perhaps as much as 85 per cent of the country's poppy production and virtually all of its heroin. The Taliban's mullahs pondered the problem of narcotics exports and announced that they were not prohibited by the Koran. Anyway, the consumers would be Christian infidels of the West, so where was the harm in supplying them with narcotics? The Taliban will have almost limitless foreign revenue when it organises the heroin trade properly.

On the night of 24 September 1996 the Taliban routed government forces in the strategic town of Sarobi, 45 miles east of Kabul and from there launched a three-pronged attack on Kabul itself, from the east, south and south-west. Government jets bombed Taliban forces, attacking a customs post three miles outside the city, but the post fell nevertheless.

Two days later Kabul was in Taliban hands. Najibullah, who had plenty of time to escape to the north, apparently believed that he would be safe in the UN compound. He was mistaken. He was dragged out of it and killed by a bullet in the head and strung up on a lamp-post outside the Presidential Palace from which his brother, Shahpur Ahmadzal, who was still alive, was also hanged and where great crowds assembled – or were forced to assemble. Noor Hakmai, a Taliban commander announced: 'We killed Najibullah because he was the murderer of our people'.

Soon after Najibullah and his brother were executed, Najibullah's bodyguard, General Jaafar and his aide, Tokhi, were hanged. The Taliban executioners exulted in the death of Najibullah and his brother. They placed a cigarette between Najibullah's fingers and pushed banknotes up the nostrils of both men. Shahpur Ahmadzai's fingers were closed around some Koranic verses which supposedly justified his murder. *Taliban Radio* announced that prayers for Najibullah were forbidden. A Taliban 'religious police' unit issued a printed statement:

> Since Najibullah was corrupt, a member of the KGB and an adulterer he did not have the right to a defence and was already condemned. Nobody is allowed to offer prayers for him anywhere in the country.

Najibullah had earned the title 'The Butcher of Kabul' and it was unlikely that anybody other than his immediate family would have wanted to pray for him.

Patrols searched for President Rabbani, Gulbuddin Hekmatyar and Ahmad Shah Massoud, all of whom were to be executed. Rabbani found refuge with General Dostam, no doubt at a price, for the two had been bitter enemies.

Soon Kabul airport and Bagram air base, 30 miles north of Kabul, were under the control of the Taliban, whose leaders announced that a six-man ruling council would administer the country. Many government troops fled to a great redoubt in the north-east through the strategic 1.6 mile Salang tunnel in the Hindu Kush. They knew that their escape route over the mountain passes would be blocked by snow by mid-October. On the night of 30 September they were approaching the southern mouth of the tunnel but by this time most of the Afghan army had passed through it.

The speed of the conquest of Kabul startled the Taliban but the rank and file were given little time to enjoy it. They were quickly sent off in pursuit of the government and its army. After a brief exchange of artillery fire, Taliban's warriors overran the Parwan provisional capital of Charikar.

Casualties during the battle for Kabul have been difficult to assess. In a battle for a military college about six-miles from the city 100 government soldiers were killed, according to a Taliban spokesman. In another of the city's regions the government claimed to have killed 150 Talibani. Probably hundreds of fighters and many more civilians were killed in the battle for Kabul but no authoritative figure will ever be known.

From Pakistan, the Afghan Islamic Press Agency declared that Afghanistan was now a 'completely Islamic state where a complete Islamic system will be enforced'. This meant that the hapless populace of Kabul would be at once subjected to all the extremes of 'pure' Islam. The few television sets and stereo systems left in Kabul were ritually smashed. Islamic patrols hunted down women who had not at once conformed to strict Islamic dress codes and beat them with rifle butts. Girls were forbidden to go to school and women were forced to give up their jobs. Thieves summarily convicted of offences had their hands cut off by meat cleavers. The Red Cross appealed to the Taliban to protect civilians and not to carry out killings but the mullahs redoubled their zealous application of the *Sharia* in the face of this 'Christian imperialist interference'.[3]

Massoud's Generalship

The Taliban started its attack against the government troops in their Panjshir Valley fortress of the Hindu Kush at 4pm on Saturday 5 October 1996. Shells and rockets, backed by fire from helicopter gunships, pounded the villages and small towns in which the troops had taken up positions. Much was at stake for Ahmed Shah Massoud, who had long prepared for this confrontation. If he could hold his ground he would destroy the myth of the Taliban's invincibility and ensure that the Islamic army could not dominate Afghanistan.

The Panjshiris, fighters for more than 2,000 years, were a formidable enemy and even the Russians never fully conquered them. Ironically, the weapons that were being used against them by the Taliban had travelled that way before. During the 1980s Soviet troops had used them to pulverise the Panjshir. Thousands of paratroopers had dropped into the Panjshir and hundreds of them had been picked off by Panjshiri snipers. At that time the Panjshiris' morale had been high but now, in October 1996, they were weary and demoralised. Massoud's options were limited but he had blocked the mouth of the valley by blasting the steep mountains on either side of the entrance and creating a rockslide. This prevented the Taliban from taking in its tanks and big guns while Massoud opened up with his own guns.

For the first time, the Taliban's opposition did not melt away and they faced a serious military confrontation. Perhaps they did not understand that Massoud was head not only of an army but a tight web of extended families, their bonds strengthened by centuries of conflict with outsiders. Even as they scrambled over the hills, taking on the entrenched Tajik defenders, they found that the Koran, which was part of the fighting gear, gained them no special military advantage here. Ordinary people elsewhere might not want to shoot a Taliban holy warrior but Massoud's Tajiks had no such compunction.

From his great base in the northern city of Mazar-e-Sharif, General Dostam, with his 20,000 Uzbek army, stood back to watch the outcome of the Massoud–Taliban struggle. His moment of decision came on 7 October when he demanded that the Taliban stop their bombardment of the government forces in the Panjshir and threatened to go to the aid of his Tajik 'compatriots' now fighting for their lives. This unbalanced Mullah Omar's calculations for he and his lieutenants knew that the combined forces of Massoud and Dostam would be powerful. Yet Dostam's move could have caused no surprise, for all Afghan leaders for centuries have been opportunistic. Only by seizing an opportunity can they survive the plotting of their many enemies. Dostam did not care about Massoud – he simply knew that should Massoud be defeated he would be the Taliban leaders' next target. The Taliban leadership hated Dostam because he was a commander in the former Communist regime. Nevertheless, his threat to the Taliban seemed to have some effect for their advance became more cautious, even while Omar was demanding that Dostam should not interfere.

Fighting between the Taliban and Dostam's forces broke out for the first time on 8 October. The place of battle was at the mouth of the Salang Tunnel, which Dostam considered Uzbek territory. He had deployed strong forces at the northern end of the tunnel and there was a small Uzbek garrison at the southern end. His overall military might is impressive and includes fighter planes and long-range weapons.

Massoud knew exactly what he was doing. Having retreated in good order to the Panjshir Valley, he allowed the Taliban to exhaust itself in fruitless attacks on positions which had withstood much more forceful Soviet assaults. Finally, in the second week of October he launched his own offensive both at the front and behind the lines, using guerrilla units to harry the retreating Taliban. His troops recaptured the towns of Golbahar, Jebel os-Siraj and Charikar, which they had lost to the Taliban at the end of September. Before long they were fighting for Bagram air base, 30 miles north of Kabul, which they captured on 19 October. At this point ex-President Rabbani demanded that the Taliban pull out of Kabul or be 'drowned in their own blood'.

Having pledged support for Massoud, General Dostam made a show of joining in the offensive. His troops appeared in Jebel os-Siraj and Charikar –

but only after the government's recapture of these places. It was widely believed that Rabbani and Massoud had offered him the post of defence minister in any fresh administration. This might have been premature because as always in Afghanistan there were wild cards, such as Gulbuddin Hekmatyar who, like Dostam, was talking with the Talibani behind the backs of Massoud and Rabbani.

Significantly, at this time the Taliban banned journalists from the front line – though any 'line' was fluid. Some commanders threatened to kill foreign journalists if they saw them again. There was trouble elsewhere for Taliban. The country's most important military airbase, at Shindad, near the western City of Herat, came under government air attack and raids by motorised units. The destruction caused by these raids was seen by journalists, some of whom penetrated the Kabul military hospital where they saw many Taliban wounded.

Meanwhile, a single-sheet anti-Taliban underground newspaper appeared in Kabul. Anybody found with it faced flogging and perhaps execution. *Kabul Radio*, now renamed *Shariat Radio*, announced that anybody found on the streets after curfew would be severely dealt with, especially foreign journalists who were 'misreporting the news'.

But they were not misreporting when they wrote that the millionaire businessman Osama Bin Laden was in Kabul and being fêted by the Taliban leadership. Bin Laden is wanted for questioning about links to the bombing of American military bases in Saudi Arabia and the US State Department describe him as 'one of the most significant financial sponsors of Islamic extremist activities in the world today'.[4]

At the same time, young men found during house-to-house searches were being given the 'chance to die for Islam' – that is, they were being drafted into the Taliban army. A Taliban terror had taken over the city and its region.

UN officers were in the field trying to find a peaceful solution to the conflict. On directions from their HQ in New York they offered a face-saving formula to the Taliban. Kabul would be demilitarised and placed out of rocket range of all factions, each of which, including the Taliban, would be allowed to open a small office in the city. It was probably too sensible a plan to hope for it to be successful, even though Massoud told the UN representatives that he had no intention of fighting his way back into the city because it had already suffered too much destruction.

Even though the Taliban had suffered a reverse at the hands of Massoud and Dostam, Russia and four Central Asian republics – Kazakhstan, Kyrgyzstan, Tajikstan and Uzbekistan – gave Mullah Omar a 'hands off warning'. Meeting in the Kazakh capital, Alma Ata, the five warned Omar that should the conflict spread beyond the border of Afghanistan into the Commonwealth of Independent States, they would take 'adequate measures'. However, in a formal communiqué the CIS group promised not to interfere in Afghanistan's internal affairs while offering humanitarian aid. Uzbekistan has, in fact, been 'interfering' for several years, in that it provides covert military support to General Dostam, an ethnic Uzbek.

Role of Pakistani Intelligence

Massoud's strategy and tactic provide an interesting lesson in warfare but then, in a negative way, so do the Taliban's operations following their capture of Kabul. According to my own sources within Pakistani intelligence (ISI) Mullah Omar and his commanders rejected Pakistani warnings not to pursue the retreating government army. In doing so they overstretched their already strained logistical organisation. Ammunition and supplies could not keep up with the ever-eager combat troops and no provision was made for fallback positions should Massoud's forces counterattack. The Taliban created their own reversal of fortunes.

Pakistan's denial of aid for the Taliban is simply unbelievable. In fact, strong Pakistani support and operational direction was crucial. The Taliban's mullahs, strong in ideological commitment, knew nothing of planning, command, control and logistics. For this they depended on the Pakistani army. The Afghan army's failure at Sarobi, on 24 September, was the result of Pakistani involvement. Intelligence officers advised the Taliban to continue a drive on the main highway while also infiltrating across the mountains. This gave Massoud no time to prepare fall-back positions on the plain east of Kabul. Meanwhile, also under Pakistani advice, the Taliban opened up their standing fronts at Rishkor and Maidanshah, producing greater difficulties for Massoud. He was reluctant to fight a battle in the rubble of Kabul and on the night of 25–26 September he abandoned the city while a rearguard on its eastern perimeter slowed the Taliban advances.

The Taliban's Structure

The undisputed spiritual leader of the Taliban is Mullah Muhammed Omar Akhund from Kandahar. Mullah Omar was a field commander at roughly battalion level in the struggle against the Soviet Union and is said to have been wounded. A 31-year old man, with only one eye, the Mullah is a mystical figure, not given to the bellicose utterances of, say, the Ayatollah Khomeini of Iran.

For Muslims, Kandahar has special significance. In 1751 the *Khiraq* (cloak) of the Prophet Muhammad was brought to the city and it is displayed to the people in times of difficulty and danger and also on occasions of great triumphs. In April 1996 the Mullah Omar held the *Khiraq* aloft for the faithful to see. On that occasion, a conclave of senior Taliban mullahs declared him Amir, or leader, of all the Muslims in the world, only the fourth such Amir to be so 'elected'. Muslims in other lands, especially the Shia Muslims of Iran, might not agree with such elevated ranking for Mullah Omar but it illustrates the purpose and drive of the Taliban.

Mullah Omar presides over a collective leadership and delegates much authority to local commanders. His deputy, Mullah Muhammad Rabbani (not related to the refugee president) led the assault on Kabul and then became its governor. For the Pakistan government, Mullah Omar is an attractive alternative to the unpopular, cruel and dictatorial Hekmatyar.

The Taliban's 50,000 warriors are divided into what would be five divisions, the nearest appropriate Western term. On the whole they are infantry but the Taliban also has an armoured wing and an artillery support group. No formal rank structure exists; the junior leaders come to prominence and command through performance in the field. Weapons and ammunition are plentiful. The basic supplies first came from Pakistan, to be augmented by vast stocks of material captured from the Afghanistan army.

The men, inspired by their Islamic zeal, are totally fearless. During their advance towards Kabul they walked across minefields under enemy fire. The prospect of death did not deter these Pashtun tribesmen; their mullahs had promised them a quick passage to the Paradise so graphically and appealingly described in the Koran. The Talibani forces are highly disciplined the result of their rigorous religious training in the *madrasahs.* They march under a plain white flag, white indicating Islamic purity.

Foreign Attitudes to the Taliban

The major powers remain profoundly interested in Afghanistan. Russia fears that its Islamic fundamentalism will cross into the Central Asian states that were once part of the Soviet Union. India has a tacit pact with Russia to contain Saudi-American influence. Strangely, considering American detestation of Islamic fundamentalism, the US government supports the Taliban; the reason could be that in ideology it is Sunni orthodox and therefore hates the Shias of Iran, the country at the top of the US blacklist. The Teheran administration of Ayatollah Rafsanjani returns the hatred of the Taliban's Sunnis, but nevertheless would do much to help them if they would take back into Afghanistan the three million refugees who now 'disfigure' Iran – to use Rafsanjani's terminology.

The situation in Afghanistan exasperates and frustrates the Iranian leadership. Having backed several groups in Afghanistan from the start of the Soviet invasion in 1979, with the intention of gaining great influence in the country, the ayatollahs have lost out to Pakistan, despite much high-level political manoeuvring. The Iranians consider that Pakistan it gaining too much power and influence in the Islamic world. India will do everything possible to undermine Pakistan's growing power in the region. It sees Afghanistan as a potential ally, once it recovers from the ruinous war, but any alliance would be impossible with a Taliban-ruled fundamentalist Islamic state.

Britain and the major European powers, while disturbed by the humanitarian crisis, are pleased that Afghanistan continues politically to preoccupy the attention of Russia, Pakistan, India and other states. Such an intense preoccupation will prevent them from causing trouble elsewhere – or so the theory runs in London, Paris and Bonn. They do not want the Taliban to become the government of Kabul but they believe that its successes could at least concentrate the minds of the various factional leaders on the need for a genuine coalition if the Taliban is to be effectively countered.

Militarily, the Taliban is a fascinating phenomenon. Its leaders had no military pretensions and certainly no professional knowledge. The rank and file knew nothing more than how to handle a firearm, and only in the most basic way; every boy growing up in Afghanistan since 1979 has learned the mechanics of shooting. While the Pakistani military provided training and guidance, generally the Taliban warriors learned their martial trade as they went along. The Taliban illustrates, yet again, how simple it is for military manipulators – in this case the Pakistanis – to turn a rabble into an 'army'.

Politically, the Taliban is a dubious experiment. Indeed, it is more reactionary than positive in its own right and its policies in 1996 were unclear. Even when militarily defeated it will not go away politically and its presence will ensure that centralised government, already virtually inconceivable, will be totally impossible. Afghanistan for the foreseeable future will become a combination of Taliban and tribal warlords ruling aggressive fiefdoms.

Religiously, the Taliban is a menace if only because its fundamentalist successes and excesses could inflame other aggrieved minorities.

In human terms, the Taliban is a disaster for it has imposed on the areas it has conquered all the cruelties of fanatical Islam and reduced every female in Afghanistan to a state of slave-bondage. The Red Cross official's prediction that a Taliban take-over would mean the end of civilisation in Afghanistan was not too great an exaggeration.

Routs and Reversals

An extraordinary series of events threw Afghanistan into even greater chaos in the middle of 1997. They began in May when General Malik, Pahlawan, second-in-command to General Dostam, defected to the Taliban, taking with him a large number of Dostam's Uzbek militiamen. The extra weight and prestige enabled the Taliban to capture three more northern provincial capitals. Mazar-e-Sharif, Dostam's stronghold, and Shibarghan, his military HQ, fell to the opposition.

The Taliban victory was a political boost to Pakistan, which had supported the group amid the wild disorder of feuding militias. Dostam fled to Turkey while Pahlawan's prize for betraying his chief was to be named as deputy foreign minister in the Taliban government in Kabul.

Neither Pakistan's satisfaction nor the Taliban's euphoria lasted long. General Pahlawan's Uzbeks mutinied after only four days of alliance with the Taliban because of the heavy-handed way in which the Taliban assumed power. The holy warriors were too extreme, too superior. In a single night in Mazar-e-Sharif the Uzbeks slaughtered hundreds, perhaps thousands of Talibani. It was a typical but extreme example of the way in which allegiances can be reversed in Afghanistan. In six days – 29 May–3 June 1997 – the Taliban retreated 150 miles, losing control of all centres in northern Afghanistan.

Meanwhile, the fortunes of Ahmed Shah Massood were rising as he advanced towards Kabul. On 20 July his veteran commander, General Bismillah Khan, captured Charikar and Bagram air base and established a front line. The Taliban still controlled two-thirds of Afghanistan. Even if the anti-Taliban alliance should succeed in forcing the Pushtun-dominated Taliban out of several provinces there is no military solution to the Afghan conflict.

Pakistan was seriously embarrassed in backing the Taliban. Also, Pakistan succeeded in embarrassing its important allies the US and Saudi Arabia who financed and tacitly acquiesced in Pakistan's intervention in Afghanistan. Losing face in Afghanistan, as Pakistan has done, takes years to recover from.

References

1. The veteran British Afghanistan-watcher, Sandy Gall, was in Kabul for *The Times* of London at this time. He reported: 'The Afghan deputy Prime Minister, Mr Ghafoorsal, told me, "Mestiri, the Tunisian special envoy for the UN, made the mistake of trying to get a consensus. Massoud is trying to get a majority of the political parties together in a coalition." ' (25 June 1996)
2. *The Times* published an editorial about the episode. It reads in part: 'For the Russians, accustomed daily to the dreary news of military bungling and disaster in Chechnya it was electrifying news. Who did not rejoice with the seven dare-devil Russian hostages at their courageous escape from captivity? It had all the ingredients of a Hollywood adventure: airmen shot down by tribal fanatics, held captive in a wartorn country, repeatedly promised freedom and disappointed as negotiations turn sour, finally triumphant after planning, pluck and luck win the day ... the captain getting the motors going one by one, the bewildered Taliban rushing onto the tarmac, the plane ducking and diving at treetop height, the hapless guards bundled up in the back of the plane, astonished but unharmed.' (19 August 1996)
3. *The Times* noted: 'For the shell-shocked inhabitants of Kabul, bombed and rocketed for the past five years in Afghanistan's bloody civil war, the lightning victory of the Taliban has brought an end to the random horror that has left thousands mutilated and reduced Kabul to rubble. That, at least is something for which they can only be grateful ... The peace they have imposed in their wake is the iron regime of the fanatic.' (1 October 1996) The residents of Kabul were 'grateful' for no longer than 24 hours. After this they were cowed, terrified, angry and resentful.
4. John Swain, *The Sunday Times.* (28 October 1996)

All major aspects of the Afghanistan War have been covered in various issues of *War Annuals, 1–7.* Those in *War Annual* 7 include:

The Mujahideen Resistance's failure to exploit advantage
The Mujahideen's attack on Gardez and the fall of Kabul
The rise of the *Taliban Farsi* and its defeat of Gulbudden Hekmatyar's forces at Charasiab.

The collapse of the Najibullah regime was described in detail in *War Annual 6.*

Background Summary

1973	King Zahir Shah ousted in military coup led by cousin Mohammad Daoud.
1978	President Daoud killed in coup. Nur Mohammad Taraki elected president of revolutionary council.
1979	Hafizullah Amin appointed Prime Minister, becomes President

	after shoot-out in which Taraki is killed. In December, Soviet troops land in Kabul. Babrak Karmal becomes President. Amin executed.
1980	UN General Assembly calls for withdrawal of foreign troops. Afghan guerrillas in Pakistan receive arms from Pakistan, US, Saudi Arabia, Egypt, Gina.
1986	Mohammed Najibullah becomes President.
1988	Pakistan, Afghanistan, the Soviet Union and the US sign agreement for withdrawal of Soviet troops.
1989	Last Soviet soldier leaves in April. Mujahideen consultive council elects moderate Sibghatullah Mojadidi President and hardliner Abdurrab Rasul Sayyaf Prime Minister.
1992	Najibullah deposed and Mojadidi becomes head of state, soon replaced by Burhanuddin Rabbani.
1993	Factional fighting leaves more than 10,000 dead.
1994	Battles reduce Kabul to rubble. Mohammad Omar Akhund sets up Taliban, Islamic student group that emerges as guerrilla force.
1995	Taliban forces reach Kabul in February, but are repelled. In October they are back.
1996	In April, a thousand Moslem clergymen choose Taliban leader Mullah Mohammed Omar as Amir, denouncing Rabbani as unfit to lead Islamic nation. On 27 September, after conquering large areas of the country, Taliban takes control of Kabul, shooting Najibullah and hanging his brother Shahpur Ahmadzal, and declaring that Afghanistan is a 'completely Islamic state'.

3

The African Great Lakes Wars

GENOCIDE IN RWANDA, BURUNDI, ZAIRE, UGANDA

There is nothing new about the enmity between the Tutsi and Hutu tribes who inhabit the central African states of Rwanda and Burundi. The mountainous Burundi is bordered by Tanzania to the east and south, Lake Tanganyika and Zaire to the south-west and Rwanda to the north. A thousand years ago Bantu peasant farmers, the Hutus, settled in the area and now comprise 85 per cent of the population of about five million. In the 17th century the cattle-owning Tutsi began moving into the country from the north and despite being a minority they soon dominated the Hutu.

As independence from Belgium approached in 1962 the Hutu and Tutsi leaders formed UPRONA (*Union pour le progrès national*) but the new organisation was unable to control ethnic tensions between the two tribes. A republic was proclaimed in 1966 and the new president, Micombero, tightened Tutsi domination of the country. In October 1971 Micombero established a military junta of 27 members but within a year the Hutus rebelled and killed probably 10,000 Tutsis. The government retaliated by slaughtering many Hutus, especially the educated. Estimates of the dead vary between 80,000 and 250,000. The modern Hutu–Tutsi wars that have so disfigured Central Africa date from this period.

In August 1988 a fresh outbreak of genocide shocked the world. The army massacred 20,000 Hutu men, women and children and tens of thousands fled to neighbouring countries.

In terms of ethnic tensions, Rwanda mirrors Burundi. With a population of nearly seven million and an area of only 26,336 sq km, Rwanda is the most densely populated country on the African continent. It is bordered by Uganda on the north, Tanzania to the east, Burundi to the south and Zaire on the west.

Rwanda's economy depends heavily on agriculture, with most of the working population growing sweet potatoes, bananas, beans, sorphum, rice and maize. Low rainfall, poor soil and the damage done by traditional methods of agriculture have made Rwanda unable to feed its population, which increased rapidly during the 1980s at an annual rate of 3.7 per cent. Over-population and under productivity have been factors in the warfare which has raged across the country.

The Hutu–Tutsi dispute is largely about land. The fertile hills and valleys of Rwanda and Burundi are uncomfortably squeezed between the forests of eastern Zaire and the dry plains of East Africa. Hutus and Tutsis have not had separate areas but have lived together for centuries on the same hills, even in the same communities. To each of them, the land is theirs – and there is nowhere else to go, at least not in peace.

The Belgians, who ruled Rwanda between 1916 and 1962, governed by indirect rule, which strengthened the power of the Tutsi minority. Tutsi domination was increasingly challenged by the Hutu majority in the run-up to independence. Indeed, the Tutsis tried to eliminate the Hutu leadership. The Hutus rebelled and much bloodshed ensued until the Belgians restored order and made democratic reforms which ensured the power of the Hutu majority.

However, in 1965 Rwanda became a one-party state and its authoritarian government was increasingly dominated by Hutus from central Rwanda. Quite apart from the Hutu–Tutsi tensions and violence, northern Hutus rebelled against the central Hutus.

Despite the many ethnic difficulties, a degree of co-operation existed between Tutsis and Hutus until 1993, but only among the better educated classes. Periodically the Tutsis massacred Hutus in their thousands but after the middle of 1993 the Hutus turned against their former masters.

For outsiders the reciprocal hatred is inexplicable because the two tribes do have links of history, culture, language and through inter-marriage. While newly arrived foreigners cannot tell the difference between people of the two tribes, the gangs of killers who roamed Rwanda in 1993–94 could so do. Fragile truces were patched together, usually with foreign intervention, but on 6 April 1994 a surface-to-air missile brought down the aircraft in which the presidents of Rwanda and Burundi were travelling. Patched-up truces were no longer possible after this.

A ferocious war began. President Juvenal Habyarimana, dictator of Rwanda for 20 years, was a Hutu and his tribe blamed the Tutsis for his 'murder'. Rwanda had two armies at the time. One was the government army commanded by General Augustin Bizimunga; the other was the army of the Rwanda Patriotic Front (RPF) of the Tutsi tribe, which attacked the barracks of the Hutu-dominated presidential guard in the capital city, Kigali.[1]

The extremist Hutu radio station, *Radio Milles Collines,* which had been used since September 1993 to build up tribal hatred, urged the killers on. The Hutu Prime Minister, the President of the Constitutional Court and many prominent political figures who might have been expected to call for peace and stability were among the first victims.

The massacres that followed were horrible in their intensity and savagery.

The International Red Cross (IRC) put the final figure – although there never is a final figure in Hutu–Tutsi relations – at 500,000 dead, mostly Tutsis. French and Belgian paratroopers evacuated several hundred foreigners from Kigali but these foreign troops would have been butchered themselves had they attempted to save the lives of Rwandans of either tribe.

Across the border in Burundi, other Tutsi–Hutu fighting went on. In October 1993, Tutsis, who dominated the Burundi army, had overthrown and killed President Melchoir Ndadaye after he had been in power for only three months. When the Tutsi coup collapsed its leaders fled from Burundi into Rwanda, where they set up a government-in-exile. In February 1994 Burundi–Tutsi soldiers from Rwanda stormed into Hutu villages in Burundi and butchered untold thousands of ordinary people. Terrified by this atrocity, an estimated 600,000 people fled across neighbouring borders, seeking a sanctuary that could only be precarious. It would be fair to say that in 1994 genocide consumed Rwanda.[2]

No Sanctuary

The cycle of murder and retaliation continued into 1995. In April, UN Secretary General Boutros-Ghali tried to raise a peacekeeping force but only a small number of 'blue helmets' were in position that month when the war in Rwanda erupted yet again. The previous year millions of Hutus had found safety in camps in Zaire and Tanzania but another 200,000 found refuge – as they thought – in Rwanda itself, in the nine camps the UN had established. But these camps also became sanctuaries for war criminals and recruiting pools for the extremist Hutu militias. The Kigali government claimed that the people of the camps were a 'threat to government authority' and urged the UN to disperse them back to their villages.

The humanitarian UN could not agree to this. Battalions of heavily-armed Tutsi government troops encircled the camps, kept out UN reinforcements and began an eviction at gunpoint. One of the worst crimes of humanity of the century then took place. At Kibeho, near the Burundi border, soldiers herded 80,000 men, women and children onto a long narrow ridge for 'processing'. Without water, latrines and medicines their plight was pitiful. Hutu militiamen among the refugees forced many men of their own tribe into a phalanx and from behind cut them with machetes, forcing them to run down the hill towards the soldiers' barbed wire fences. Here the Tutsi troops shot them down. The Hutu militiamen, prodding and slashing, created panic among the desperate Hutus who broke through the Tutsi barriers. Here the troops opened up with machine-guns and even rocket-propelled grenades (RPGs) – against inoffensive women and children! The small number of UN peacekeepers present could only watch in anguished frustration.[3]

At the UN Security Council's request several countries have sent peacekeeping forces, many medical teams and experts in water purification to Rwanda and vast amounts of money have been spent in aid. However, outside diplomatic intervention, which was well meant, had the effect of setting off the bloodiest massacres. The UN organised a campaign to secure power-sharing for Rwanda's Tutsi minority. This provoked extremist Hutu politicians into working out a plan to rid the country of all Tutsis, even the very young. To incite peasants into murdering their neighbours the Hutu leaders exploited the historical fears among Hutus of a return to Tutsi hegemony.

During 1996 and 1997 Hutu extremists embarked on a series of atrocities against expatriates working in Rwanda in a deliberate effort to scare away foreigners. One of the worst attacks took place at Ruhengeri, 135 miles northwest of Kigali, on 18 January 1997. Three Spanish aid workers were shot through the head, an American escaped death but needed to have his leg amputated and three Rwandan soldiers who tried to defend the aid compound were also killed. This atrocity was the latest of many attacks on aid workers, hospitals and foreigners, which multiplied after the return of more than 600,000 Hutus from eastern Zaire at the end of 1996.

The week before the Ruhengeri killings, Hutu extremists murdered a prosecution witness, her husband and seven children after she appeared before the UN trials in Arusha, Tanzania. She had been promised protection from killers intent on silencing witnesses to the genocide in 1994.

The Mokoto Massacre

Militiamen and members of the defeated Hutu army, furious at losing their war against Tutsi rebels, joined with their Zairean Hutu brethren and began killing Zairean Tutsis. Their objective was a 'Hutuland' on Zairean soil that would provide a safe haven for refugees and a base for armed incursions into Rwanda.

The chaos in Eastern Zaire illustrates the distrust and hatred infecting a region traumatised by the civil wars in Rwanda and Burundi. No sanctuary exists for the people fleeing from the machete-wielding warriors. In May 1996 about 800 Tutsis fled from their homes in the Masisi Highlands and took refuge in a church in the grounds of the Trappist monastery in Mokoto. The Brothers tried to protect them from the onrushing Hutu militiamen who did not at first enter the church. But in a panic some of the desperate people ran through a rear door to hide in the bush. Here they were hunted down. A Hutu cut off a woman's hands and feet. The killers cut out a man's heart. A French Trappist brother, Victor Bordeau, picked up a baby, still breathing but drenched in its mother's blood. Some monks had already fled and it was time for Bordeau and a few others to flee in the monastery vehicle before they too were butchered. As they raced away they saw the Hutu killers storm into the church and monastery where hundreds of Tutsis had imagined they might find safety.[4]

Arms for the Asking

Arms were reaching Rwanda throughout the troubles of 1994–95, mainly through Zaire. The UN Secretary General Boutros-Ghali, failed to gain permis-

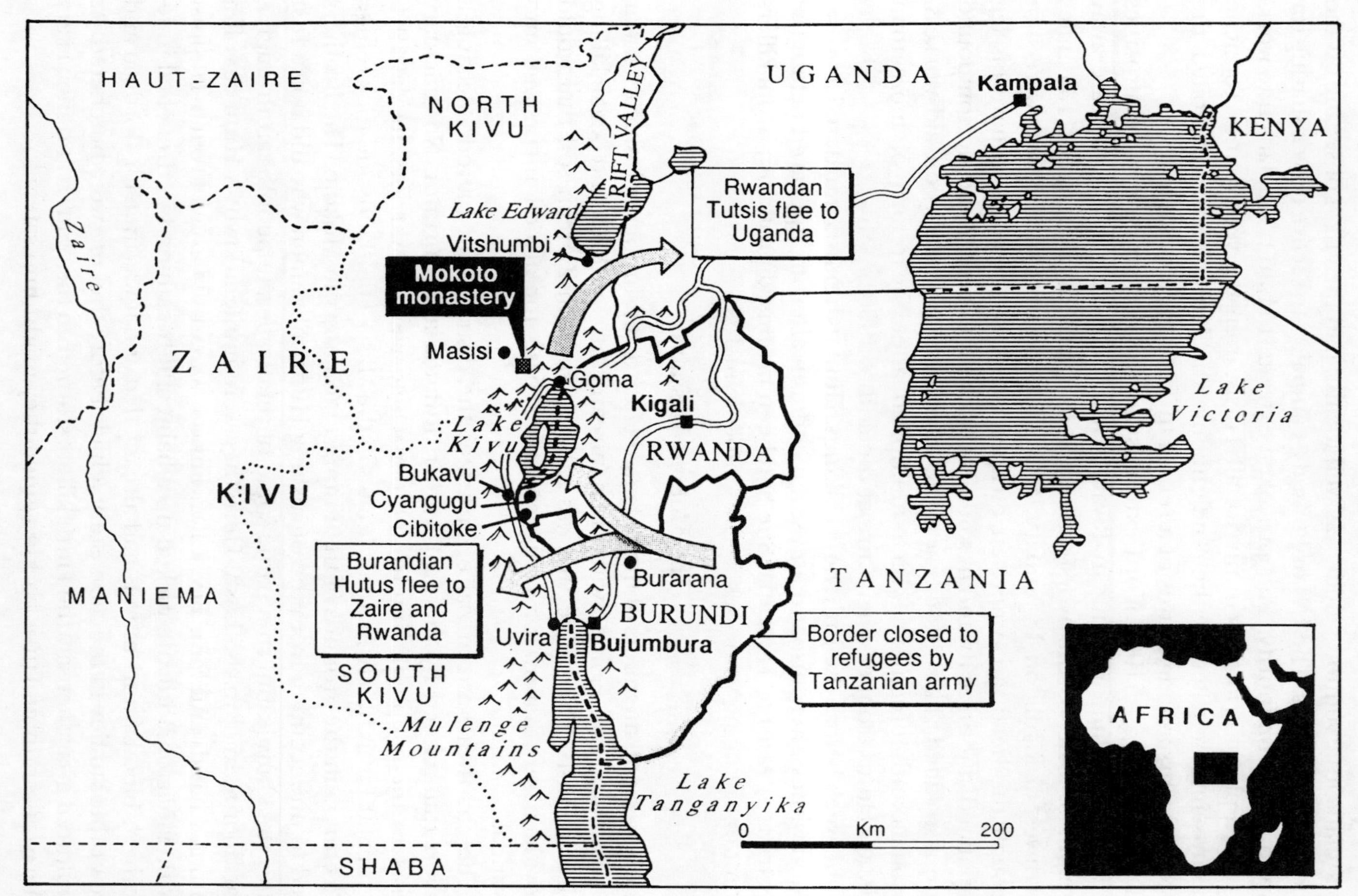

African Great Lakes Wars

sion to post UN observers at airfields and transit points in Zaire to monitor arms movements. According to Arms Protect of Human Rights Watch, France, China and South Africa had discussed or prepared arms deals with the former Rwandan army. France and China denied this while sources in Pretoria said that 'elements of the apartheid-era' might have struck deals with former allies in Rwanda.

Late in November 1996 the British Customs and Excise Department began an investigation into allegations that a UK firm had supplied Hutu militias with arms during and after the 1994 massacres in Rwanda, in breach of a UN arms embargo. The BBC and *The Times* reported on 18 November 1996 that the British Mil-Tec Corporation, based on the Isle of Man, had supplied the Hutu militias and the Rwandan government army with mortars, assault rifles, ammunition, grenades, rockets, rocket-launchers and explosives worth more than £3.3 million between 17 April and 13 July 1994. The reports were based on documents abandoned the previous week in the commune of Sake, Eastern Zaire, by the Hutu militia and the former Rwandan army and picked up by foreign aid workers.

Zaire's Role in the War

Zaire has had as corrupt a government as any on earth and President Mobutu, in power from 1965–97, was one of the world's most self-serving rulers. With this background it is not surprising that Zaire has failed to play a neutral humanitarian role in the Rwandan wars.

During 1996 Hutu refugees from Rwanda lived in vast camps in Zaire (and Tanzania). Concealed among them were the remnants of the old Rwandan army, whose middle-ranking leaders turned them into a guerrilla force. From the camps, these ex-soldiers, backed by other Hutu militias, raided across the border, attacking installations of the new Rwanda government. The Zairean government made no effort to stop them, disarm them or remove them from the civil camps.

Local politicians saw an opportunity of exploiting the wretched Rwandans by inciting the rivalry between East Zaire's 'original' inhabitants and the Tutsis and Hutus who came later. (Here it is necessary to explain that 'later' means several hundred years later!) The closely integrated Hutus and Tutsis are known collectively as the Banyarwanda and they have been the majority since 1910, though political power has remained in the hands of the local Zaireans.

The Mobutu regime treated the Banyarwanda people as foreigners and would not give them citizenship. In 1995 they even lost property rights. From time to time Mobutu and his gang, acting as agents provocateurs, created wars between the Banyarwanda and the Zaireans.

With the influx of more than one million Rwandan Hutus in 1994–96 the numerical balance of power moved, so the Zairean national and regional governments encouraged their people to attack the Banyarwanda. As if this were not enough savagery, many Banyarwanda people, now professing distinctive Hutu or Tutsi allegiance, fought each other. Further to compound the misery of the Hutu refugees, the *Maji Maji Ingilima*, a Zairean militia, attacked Hutus in settlements and camps.

There was yet a further dimension to the fighting in 1996 and 1997. A Tutsi tribe, the Banyamulenge, had been living peacefully for generations in the Mulenge Mountains of South Kivu but in September 1996 the Zairean army attacked them, killing many unarmed people. Their plight was made worse on 7 October when the South Kivu deputy governor ordered all 450,000 Banyarmulenge to leave the region. Otherwise, he said, they would be 'hunted down as rebels'.

Attacking the Banyamulenge was not only a wilful atrocity but a strategic error. These people live in mountains with few roads and, as bush guerrillas, the Banyamulenge easily ambushed the national army. The army made up for its lack of success against the Banyamulenge fighting men by committing atrocities against their women and children in the villages.

Why did the Zaire government commence this conflict? Its spokesman in Kinshasa said that Rwanda had enlisted several thousand Banyamulenge into its army, trained them and then infiltrated them back into Zaire to destabilise the eastern region and create a new Tutsi state. This is entirely possible, though the Rwandan government denied the charge.

Roberto Garreton, the UN Special Rapporteur on Human Rights in Zaire, several times accused Zaire of 'government-sponsored massacres'. This, too, is entirely believable. Zaire was in such a state of confusion during Mobutu's absence in Europe for medical treatment that army units were fighting one another, generally on a tribal basis.

The principal losers in the Eastern Zairean conflict were the Banyarwanda, who saw their homes and land stolen.

The Battle of Goma

By November 1996 Goma, the capital of eastern Zaire and the humanitarian lifeline for hundreds of thousands of Rwandan refugees, was the last major town along Zaire's borders with Rwanda and Burundi not yet completely in the hands of Tutsi rebels and Rwandan forces. On the first day of that month Goma was attacked. In a well-organised, well-armed operation, 15,000 men of the Rwandan army invaded the town by land and from a number of small boats that crossed Lake Kivu into Zaire. Goma came under shellfire, while in the city the two sides fought with mortars, AK-47s and rocket-propelled grenades.

Rebel Tutsis, backed by the Rwandan army, had already occupied a very large portion of South Kivu, from Lake Tanganyika north to Lake Kivu. In North Kivu, at this time, rebel Tutsis controlled a 100-mile stretch along the Rwandan border with Uganda. The 'allies' were also succeeding in pushing Rwandan Hutu refugees deeper into Zaire.

The regional Tutsi rebel leader, Laurent Desire Kaliba, announced that the 'uprising' was not ethnic but a political protest against President Mobutu. It had begun, said Kaliba, a few days before, when the Zairean authorities had ordered ethnic Tutsis off the land they had farmed for generations. At that time, Tutsi fighters, armed and trained by the Tutsi-led Rwandan army, had fought with great determination and skill to block and then drive back government forces trying to save the key towns of Uvira and Bukavu.

Some Hutu residents of Goma panicked and joined thousands of other Rwandan Hutu refugees fleeing the area. Gangs of looters, which according to aid workers included Zairean soldiers, ransacked Goma.

As so often in the Great Lakes wars, aid workers and missionaries were in grave peril. About 100 foreigners were trapped in a cathedral in Bukavu. These and another 100 aid workers were gathered in UN compounds for evacuation, which eventually took place.

Zaire acted decisively in the Rwandan conflict on 21 January 1997, when it actually declared war against the Tutsi-backed rebels in Eastern Zaire. President Mobutu approved the declaration of war before leaving Kinshasa for further cancer treatment in France. It was then left to his Prime Minister, Leon Kengo, formally to spell out the government's intention. He declared:

> In the name of the government and with the agreement of the president of the republic, I have today ordered our armed forces to conduct the war and recover regions currently under occupation and restore the authority of the state.[5]

The government had been saying for weeks that it would launch a counter-offensive. The rebels, led by Laurent Kabila, an old enemy of Mobutu, took up arms against the government in September 1996, after Rwandan Hutu refugees and Zairean officials threatened to expel local Zairean Tutsis. The reason given was that they 'were not true Zaireans'.

With the backing of government in Rwanda, Uganda and Burundi, the Zairean dissidents took advantage of the confusion and drove the demoralised and unpaid Zairean soldiers from a 400-mile strip along Zaire's eastern border. Capturing town after town, Kabila's army inflicted a personal defeat on Mobutu when they captured his own gold mine and the north-eastern trading centre of Bunia, also regarded as Mobutu's personal property.

Prime Minister Kengo tried to rally his citizens. In a broadcast he called for them 'to stand together with our armed forces'. He said the government would continue to repatriate the Rwandan refugees who had been living for three years in eastern Zaire. 'The concern of our country has always been to contribute to the return of the refugees in dignity and in peace', he said, 'according to the wishes of international community.'

This was the first occasion on which a Zairean politician had spoken sympathetically about refugees but there was no prospect that Kengo's words would be followed by 'dignified' action.

Burundi: 1,000 Murders a Month

By 1996 it had become traditional and customary for the Tutsi-dominated army and the Hutu militias to do bloody combat with each other. This resulted in a degree of ethnic cleansing comparable only to that in Bosnia-Herzegovina. It may be simply expressed: the Tutsis drove the Hutus from the towns while the Hutus forced the Tutsis out of the rural areas. In the whirling cycle of raids and counterattacks, by June 1996 more than 1,000 people a month were being murdered, the great majority of them older men, women and children. They were much easier to attack.

President Sylvestre Nibantunganya, whose two predecessors had been assassinated, found sanctuary in the American Embassy in Bujumbura. On 22 July 1996, Pierre Buyoya took power and stifled any movement towards democracy. His coup was at once condemned throughout the world but Buyoya made his position and that of his brother Tutsi officers very clear. 'A lack of strictness is what has brought us to this point', he said 'and we are not going to allow it to go on. We have been subjected to genocide but now the army will be able to combat the killers effectively.'

There was truth in what he said. Burundi's Tutsi minority – 15 per cent of the population – had been much reduced by massacres at the hands of the Hutus. By July 1996, the month of Buyoya's seizure of power, about 200,000 Burundians on both sides had been killed since 1993. The UN and the Organisation of African Unity (OAU) called for a peacekeeping force to be sent in but the West was reluctant to commit troops to such a dangerous country. The only countries likely to contribute peacekeeping troops were Uganda, Tanzania and Ethiopia but it would be almost impossible to find troops of any nation acceptable to both Tutsis and Hutus.

Other African nations applied sanctions to Burundi, bringing it close to bankruptcy and forcing Buyoya to recognise political parties. He even restored parliament, in which the Frodebui Party, which is Hutu-backed, has a majority. The Parliament met only once, in September 1996, but nothing could be achieved because of the intimidation of the Frodebu, 22 of whose members, out of the 81 parliamentarians, had been murdered.

Meanwhile various groups armed themselves still further, notably the National Council for the Defence of Democracy (CNDD). The CNDD's guerrilla militia, using weapons supplied by or through Tanzania, may be pivotal in the conflict. The role of Tanzania itself will also be crucial because it set up, in 1996, a large training camp for the Hutu militia.

During 1996 and well into 1997, the Hutu guerrillas carried out many operations, such as repeatedly cutting Bujambura's power lines. It has become so dangerous to travel on Burundi's roads that all foreign visitors fly in and out.

THE MASTERMIND OF GENOCIDE?

The names of the men responsible for organising the horrendous massacres of 1994 did not at first come to light, perhaps because so many people, though interested in seeing justice done, believed that the slaughter was 'simply' another manifestation of the ancient hatreds between Tutsis and Hutus. However, between March and June 1996 about 15 known Hutu extremists who fled Rwanda after the war were arrested in Cameroon.

The most important man by far among those being held was retired colonel Theoniste Basogora, aged 55, from the Hutu-dominated north-western province of Gisenyi. According to human rights monitors, Basogora was the organiser of 'Zero Network', a group of death squads which commenced a campaign of massacre late in 1991. Belgian and French intelligence sources, who know more about this region than anybody else, point out that Basogora was a close associate of the wife of President Juvenal Habyarimana and of her three Brothers. It is likely, say some of these sources, that these four people had the original idea of murdering Tutsi civilians. The purpose was to eliminate all potential support for a Tutsi-led insurgency.

Talks of reconciliation between Hutus and Tutsis took place in Arusha, Tanzania, in 1992 and Basogora opposed everybody who wanted harmony. He specially targeted Prime Minister Agathe Unwilingiimana, a leading Hutu moderate. By no coincidence she was one of the first leaders of the Hutu moderates to be murdered – or 'executed' as Basogora said – in 1994.

Having caused so much trouble within the government and the army that he was retired, the vindictive Basogora now established radio contact with three battalion commanders in Kigali, one of whom commanded the presidential guard, which is known to have murdered 10 Belgian peacekeepers.

A UN report published in May 1996 stated that Basogora used a South African middleman to ship weapons from South Africa through the Seychelles to Goma, Zaire. This violated a UN arms embargo. Belgium, the first to request Basogora's arrest, wants him tried for the murder of its 10 peacekeepers. The International Criminal Tribunal for Rwanda, based in Arusha, wants to see him in court on a charge of genocide. Rwanda's Tutsi-led government in June insisted that it had first priority in dealing with Basogora. Basogora is defended by an able Belgian lawyer, Luc Temmerman, who protests that his client was merely acting to protect his country from Tutsi invaders.

Any trial would last a long time, unless Basogora's principal aides, also in custody, can be persuaded to give evidence against him.

RETURN OF THE MERCENARIES

During the 1960s white mercenaries fought in central Africa, notably in the old Belgian Congo, which became Zaire. There were too few of them to be effective as a military force but they became notorious and newsworthy. In 1996 agents for President Mobutu recruited another legion of mercenaries, soldiers of fortune from France, Belgium, Britain, Eire, Germany, Croatia and Serbia. They were paid directly by Mobutu but they came under the Zaire government, which announced in January 1997 that these men were in the country only as instructors.

Indeed, they were instructing – the Zairean army needed all the professional help it could get – but they were also leading soldiers in efforts to regain territory in the east of the country, that is, in Kivu province.

In January, some mercenaries were flying Mi24 helicopter gunships, dating from the days when the Soviet Union supplied Zaire with military equipment. In this high-profile role the mercenaries were expected to raise the morale of the army, which is underfed, under-equipped and rarely paid. The mercenary fliers were mostly from Eastern European countries, a fact conceded late in January 1997 by the newly-appointed chief-of-staff of the army, General Mahele Lieko Bokungo.

Even in large numbers infantry mercenaries could not hope to make inroads against the trained guerrillas of the Alliance of Democratic forces for the Liberation of Congo-Zaire, led by Laurent Kabila. However, if Mobutu and his military leaders could buy enough experienced helicopter and jet pilots they could just possibly defeat the rebels in North and South Kivu provinces. The rebels possess no anti-aircraft weapons.

The presence of mercenaries further serves to indicate the chaos brought about by the Great Lakes wars. They are rumoured to be receiving between $1,000 a month for infantry and $4,000 for pilots.

TENSIONS BETWEEN BURUNDI AND TANZANIA

In December 1996 the fear of yet another war alarmed observers of the Central African chaos. Diplomats in Bujumbura warned their governments that Burundi's overwhelmingly Tutsi army was about to invade Tanzania in order to attack Hutu rebel bases there. At the time these bases were grossly expanded with fighters driven out of Zaire. The aim of the strike into Tanzania, the diplomats said, would be to prevent the largest Hutu rebel group from establishing a foothold after being routed in Zaire.

The rebel group is the National Council for the Defence of Democracy (NCDD) led by Burundi's former interior minister, Leonard Nyangoma. Burundi's military leader had accused the Tanzanian government of turning a blind eye to the growing NCDD presence on its soil, but Tanzania has consistently denied that it has given anything more than humanitarian aid to the rebel organisation. Nevertheless, hard-headed foreign diplomats have always known where Tanzania's sympathies lie. Julius Nyerere and other Tanzanian leaders have long tolerated Hutu rebel groups, even the overtly extremist forces such as the *Interahamwe* and the *Palipahutu*, who talk of their 'work' – meaning the killing of Tutsi.

All sides to the Great Lakes conflict are blindly shackled to the primitive principle, 'If we don't strike first, we risk annihilation'. With this as a virtual creed the killing can only continue.

UGANDA CIVIL WAR: ANOTHER CONFLICT IN THE NAME OF THE LORD

Background Summary

After the brutal Idi Amin was overthrown in 1980, Milton Obote became President of Uganda but his Defence Minister, Yoweri Museveni, claimed that the election had been rigged (which the UN accepts that it was) and took to the bush to wage a guerrilla war against him. He gathered about 6,000 fighters, created the National Resistance Army (NRA), and was armed and supplied by Libya.

The Ugandan army, while pursuing the NRA, committed countless atrocities against civilians, butchering at least 300,000 of them. In July 1985 Brigadier Okello overthrew Obote and asked General Tito Okello (no relation) to become president, but the army remained out of control. Museveni, having strengthened his NRA, moved out of his own area in the south-east, known as The Triangle, and promised to restore order to Uganda. On 26 January 1986 he captured Kampala, deposed Okello and declared himself President.

Peace was elusive. General Basilio Okello (no relation to the two other Okellos) raised an army from the great Acholi warrior tribe to fight Museveni, while men from the disbanded armies of Obote and Tito Okello formed the Uganda National Liberation Army (UNLA) to oppose Museveni and his NRA, which was now, in effect, the new national-army.

On 18 January 1987, the NRA won a significant seven-hour battle at Corner Kilak, defeating the 'Holy Battalion' of the UNLA. Museveni achieved a degree of peace and stability and showed that he had political will when he disbanded some mutinous regiments. About 20,000 rebel soldiers accepted an amnesty offer in 1988 but Museveni still faced serious security problems in the north and east. Museveni claimed to be a friend of the Western democratic world and he solicited American economic and military support. Even so, he was prepared to employ soldiers from North Korea, East Germany, Cuba and Libya, all countries opposed to the West. He was also allowing Uganda to be used as the site for Libyan-financed bases and military training camps to accommodate men opposed to President Mobutu of Zaire and Daniel Arap Moi of Kenya.

Museveni told Western ambassadors in 1988 that he was trying to close down camps for foreign dissidents and that he was distancing himself from President Gaddafi of Libya. At the time, neither claim was true.

Uganda's witchcraft cults caused Museveni great problems, especially the Holy Spirit Movement (HSM) founded by Alice Lakwena, a 27-year-old priestess and herbalist. She managed to form an army of ex-soldiers and peasants and assured the men that when smeared with her 'magic oils', they would be invulnerable to bullets and exploding shells. About 7,000 of these warriors were killed when they charged NRA positions and Alice fled to Kenya. Another influential cult leader was Joseph Kony, cousin of Lakwena, who taught his rebels to go into battle shouting 'James Bond! James Bond'. The vast majority had no idea what the words meant. Kony personally executed those of his men who displeased him and starved other 'offenders' by telling them that if they took food a snake would bite them on the mouth. Yet another group, the 'Red Commodores' hacked their enemies to death with machetes.

The War in 1996–97
AIDS and Private Armies

In the 1990s Uganda was struggling with one of the world's worst AIDS epidemics – with the Army especially hard hit – overspending on defence, social instability and the menace posed by the many private fictionalised armies. In 1996, under Major General Salim Saleh, the President's half-brother, the security services were gaining ground in the wars against the ruthless rebel armies. Museveni had appointed Saleh as 'Presidential Special Advisor on Military and Political Affairs' – mainly in northern Uganda – and Saleh, a popular officer, at once asked for approval for soldiers to shoot rebel leaders on sight.

The titular head of the Department of Defence and of the Uganda People's Defence Force (UPDF) is General Amama Mbabazi while the Chief-of-Staff is Brigadier Chefe Ali, but it is Salim Saleh who has Museveni's ear. Confronting Saleh in 1996–97 were these main armed rebel groups:

- The Lord's Resistance Army (LRA) led by the brutally coercive Joseph Kony. If he had his way Uganda would be ruled under a fundamentalist interpretation of the Ten Commandments and Kony has made it clear that Uganda's many Muslims 'must be prepared to embrace Christianity'. The LRA recruits among the Acholis in Uganda's far north.
- West Nile Bank Front (WNBF) led by Colonel Juma Oris. A Muslim from Uganda's north-west region, Oris and his colleagues want to restore Idi Amin to power. Under his dictatorship they had power, prestige and affluence, but Museveni reduced them to the level of middle-ranking civil servants. Sudan is a safe haven for the WNBF. Amin, comfortably exiled in Saudi Arabia, would no doubt like to return to power but General Saleh would certainly have him shot on sight.
- Uganda National Resistance Front II (UNRF II), founded by another Muslim, Colonel Ali Bamboos. Both the WNBF and the UNRF II are prepared to fight the government army and the LRA. Bamboos has no particular affiliation with Idi Amin, except that both are Muslims.
- National Democratic Alliance (NDA), under a disaffected major, Harboured Itongwa. The NDA's safe haven is Kenya, where its men were trained until August 1996. Except in an official sense, this has not

changed. The long Uganda-Kenya border cannot possibly be sealed.
- National Army for the Liberation of Uganda (NALU), led jointly by two former army officers from the Uganda-Zaire border.

All the factional armies combined – an impossible concept – could not defeat the NRA but they can keep the country in a constant state of tension and unrest. Outside intervention is also damaging. The Museveni government accuses the Sudanese government of supplying the LRA by parachute and by landings at remote airfields. It might seem odd that Sudan's extremist Islamic government would help extremist Christians but there is a reason behind the apparent madness. President al-Bashir of Sudan did a deal with Joseph Kony of the LRA: Sudan would give Kony arms if he would attack the Sudan People's Liberation Army (SPLA) of Colonel John Garang, and this Kony has done.

The LRA is probably the biggest internal problem for Museveni and Salim Saleh because its camps are so difficult to find. The discovery of four of them in June 1996 was a short-lived triumph as Kony's men melted into the bush to take up new positions in the large land.

The government set itself some interesting military targets for the end of 1997. It would create and train a rapid reaction force and acquire several more helicopter gunships to augment its 1996 fleet of only four. Mercenaries were to be imported to train Uganda specialists in helicopter gunship support of élite infantry.

New Leader in Zaire – now Democratic Republic of Congo

During fighting that went on for much of 1997, Laurent Désiré Kabila and his troops gradually overcame the forces of President Mobutu, and as early as May Kabila could claim three-quarters of the country as his own. He had not at that time called himself 'President' but he functioned as one. Either he or his representatives were negotiating with foreign mining companies; eager to exploit Zaire's massive reserves of diamonds, copper cobalt and zinc. Kabila was flying about his large country in a luxury jet, the gift of a foreign mining company. However, his aides said that Kabila was selling the nation's wealth for the benefit of the people, unlike the grotesquely corrupt Mobutu who amassed vast fortunes for himself.

In May the new regime established its offices at Lubumbashi, 1,500km to the southeast of the capital Kinshasa and here, alarmingly, many deals were being made in private. Observers reported that Kabila had already squandered much international goodwill over the massacre of Hutu refugees near Kisangani when he provided that he was unable (or unwilling) to control the killers responsible. He also engaged in an ill-advised war of words with the UN Secretary General Kofi Annan, and he set a rigid 60-day deadline for the repatriation of the Hutu refugees to Rwanda. To get Kabila to attend talks with Mobutu, Nelson Mandela, South Africa's president, in the role of peacebroker, bluntly said that Kabila's failure to do so would be a personal affront. Kabila gave way. He had been a successful general in the war against Mobutu but late in 1997 diplomats with experience in the region told me that he would turn out to be 'just another African self-seeker'.

Meanwhile, black days had fallen on Rwanda yet again. After November 1996, Hutu refugees began returning home in their thousands. Jubilantly, President Pasteur Bizimungu toured the border-lands and proclaimed to the Hutus, 'This is an historic moment for Rwanda. You are our brothers'. Unfortunately, among the returning refugees were thousands of the perpetrators of the 1994 genocide. They established camps in the country's northwest, to make a last stand. From here they send out killer gangs to eliminate witnesses to their crimes.

In addition, massacres were occurring frequently, sometimes several times a week. One night 14 schoolgirls were murdered in their dormitory along with their Belgian teacher. Buses were stopped on rural roads and the passengers slaughtered. UN observers said that it was generally impossible to know which side was responsible for an attack – the ringleaders of the genocide or government troops. The soldiers are often ineffectual. In June, rebels attacked a gaol and released 120 of their comrades. This still leaves more than 100,000 genocides suspects in filthy gaols. More than 30,000 had not, by August 1997, been charged with any crime. Trials had begun early in the year but, again by August, fewer than 150 had taken place. This was partly because families and friends of the imprisoned men had killed most of the lawyers.

It is now just possible that the killing, on and off the battlefield, will be greatly reduced, or even stopped. The rebels no longer have access to supply routes from the Democratic Republic of Congo, formerly Zaire, and by September 1997 they were running in the Zairean civil war were returning to Rwanda, to be sent at once to the northwest to fight the genocidal rebels.

As in so many countries, hatred will keep the conflict going. In Rwanda, bullets are not really essential; the machete is the preferred weapon.

References

1. The RPF was no raggle-taggle army but trained and, by African standards, disciplined. As child refugees, these Tutsis had grown up in Uganda where they joined the rebel movement of General Yoweri Museveni, himself a Tutsi. When his army took over in Uganda in 1986 it was largely led by Rwandan Tutsis. These leaders plotted their return to Rwanda and in 1990, stealing Ugandan weapons, they formed the RPF. Museveni is seen by his many enemies, at home and abroad, as a supporter of Tutsi regimes. This could make him a target for extremists. While Uganda is notable in the late 1990s for political stability, a guerrilla war was in progress in 1997 in the north. If Rwanda and Burundi were to collapse again the adverse effects would be sharply felt in Uganda.
2. The British Foreign Office was certain that the massacres did not 'just happen'. A FCO briefing paper *Problems and Progress in Rwanda,* of March 1996, stated: 'The minority Tutsi group was the target of a well-planned premeditated genocide by Hutu extremists, while moderate Hutu politicians and their supporters were killed in simultaneous massacres.'
3. One of the officers told me that he believed 5,000 people were butchered, possibly one third of them by the Hutu militiamen. The UN's representative in Rwanda gave the number of dead

as 2,000. It might seem strange that Hutus kill Hutus but in Rwanda, Burundi and Zaire this frequently happens. The aggressors are the extremists, who are prepared to kill their own people if by doing so they can gain a victory over the Tutsis. Also, many ordinary, peaceable Hutus are prepared to live alongside Tutsis. The extremist elements kill people 'guilty' of such 'betrayal'. Finally, there is animosity between Hutus from various regions and it is sometimes fierce enough to lead to bloodshed.

4. This testimony from Brother Bordeau, who related his story to a *Time* magazine reporter, Andrew Purvis (*Time*, 8 July 1996) and later to the regional representative of the United Nations High Commissioner for Refugees.
5. One report stated that up to 120,000 Hutus had moved deeper into Zaire to set up a base from which to attack Rwanda. Other reports suggested that the Hutus intended to turn the region, which is around Masisi, into a Hutu homeland.

4

Algerian Civil War: A Totally Mad Conflict

HOW IT ALL BEGAN

Algeria began its slide into civil war and chaos at the end of the 1980s after a decade of many strikes and riots, all bloodily put down by the security forces. In October 1988 President Benjedid Chadli declared a state of emergency and promised sweeping reforms to the constitution. In the event, reform was limited; no new political parties could be formed though independent candidates could fight elections. After fierce opposition, the government backed down; even so, new parties could not be based solely on religious, professional or regional interests. There was no hope whatever that this limitation would be observed, especially in a country rife with Islamic revolutionary fervour.

Extremist Muslim gangs caused havoc to Algeria's social fabric. Most fanatical was the *Front Islamique de Salut,* known as FIS or Islamic National Front. In mosques all over the country, the FIS built up a nationwide network and won over the urban poor and unemployed as well as many dissatisfied young men. In April 1990 FIS crowds demonstrated in Algiers, violently demanding dissolution of the People's National Assembly and the introduction of the *Shari'a,* the extremist Muslim code of common law.

In June 1990, in local elections, the FIS gained control of 32 of Algeria's 48 provinces and 853 of the 1,153 municipalities. This amazing result alarmed Chadli and his government. They announced fresh economic and welfare policies but at the same time restricted electoral campaigning in the mosques and increased the number of constituencies in order to swing the electoral process against the FIS and in favour of the *Front de Libération Nationale* (FLN), which had ruled Algeria since independence from France after the atrocious war of 1954–62.

The consequent FIS protest was predictably violent. The government declared a state of siege, suspended elections and swamped Algiers with troops. The main leaders of FIS, Abbasi Madani and Ali Belhadj, were arrested, as well as thousands of FIS 'traitors, ringleaders and enemies of the state'. From this moment the FIS regarded itself at war with the government.

The government could no longer put off the general election and the presidential election but barred the gaoled FIS leaders from standing. The FIS leadership was inclined to boycott the election but the highly politicised Iranian mullahs who were advising them urged them to stand. On 26 October 1991 the FIS won 188 seats; the Berber parties 25 seats, independents 3 and the FLN, the ruling party, a mere 15. In the second round on 16 January 1992 the FIS gained the additional 28 seats that it needed, under the constitution, to win outright.

The FIS would therefore form the government and Algeria would at once become an Iranian-style Islamic state, with all its attendant cruelties and retrogressive policies. Shocked and politically weakened, President Chadli announced that he would share power with an FIS government but the army chiefs refused to accept this compromise and forced Chadli to resign. On 13 January 1992 the Higher Security Council took power and brought into being a High Council of State, whose president was Muhammad Boudiaf, a hero of the War of Independence who had been in exile since 1964.

He survived only until 29 June when a member of his own bodyguard, an undercover FIS killer, assassinated him. The High Council of State appointed Ali Kafi, leader of the Algerian Veterans Association, to succeed Boudiaf and Abdesselem Belaid became prime minister. The effective ruler in 1993 was the Minister of Defence, General Nazzar.

Assassinations, bombings, massacres and kidnappings became commonplace. Many of the atrocities were carried out by the *Group Islamique Armée* (GIA). Algerian journalists, foreign civilians and missionaries were specially targeted. The GIA butchered four Roman Catholic priests in retaliation for the deaths, at the hands of French commandos, of a GIA gang that had hijacked an *Air France* aircraft.

A new president, Liamine Zeroual, appointed in January 1994, achieved no more than his predecessors and the war became more intense. At the end of 1995 General Said Bey, in a 10-day operation, ambushed GIA terrorists in the rugged country around Ain Defla and 'eradicated' 650 militants. This did nothing to break the spirit of the FIS, which by now considered that anybody connected with the 'infidel state' – no matter how tenuous the connection – was legitimate target. This even included the children of officials.

While the government's intelligence system had penetrated the FIS and GIA, the 175,000-strong security forces, backed by a larger number of helicopter gunships, was not able to contain terrorist activity, let alone defeat the Islamic militants. The government has had strong support from France and from Europe generally, but even this has had no apparent effect in overcoming the FIS/GIA, which in many areas operate at will. By the end of 1995 more than 50,000 people had been killed in Algeria.

The War in 1996–97

Given the atrocity-reprisal-atrocity mentality of Algeria's government and military on the one side and the FIS/GIA on the other, it was always inevitable that

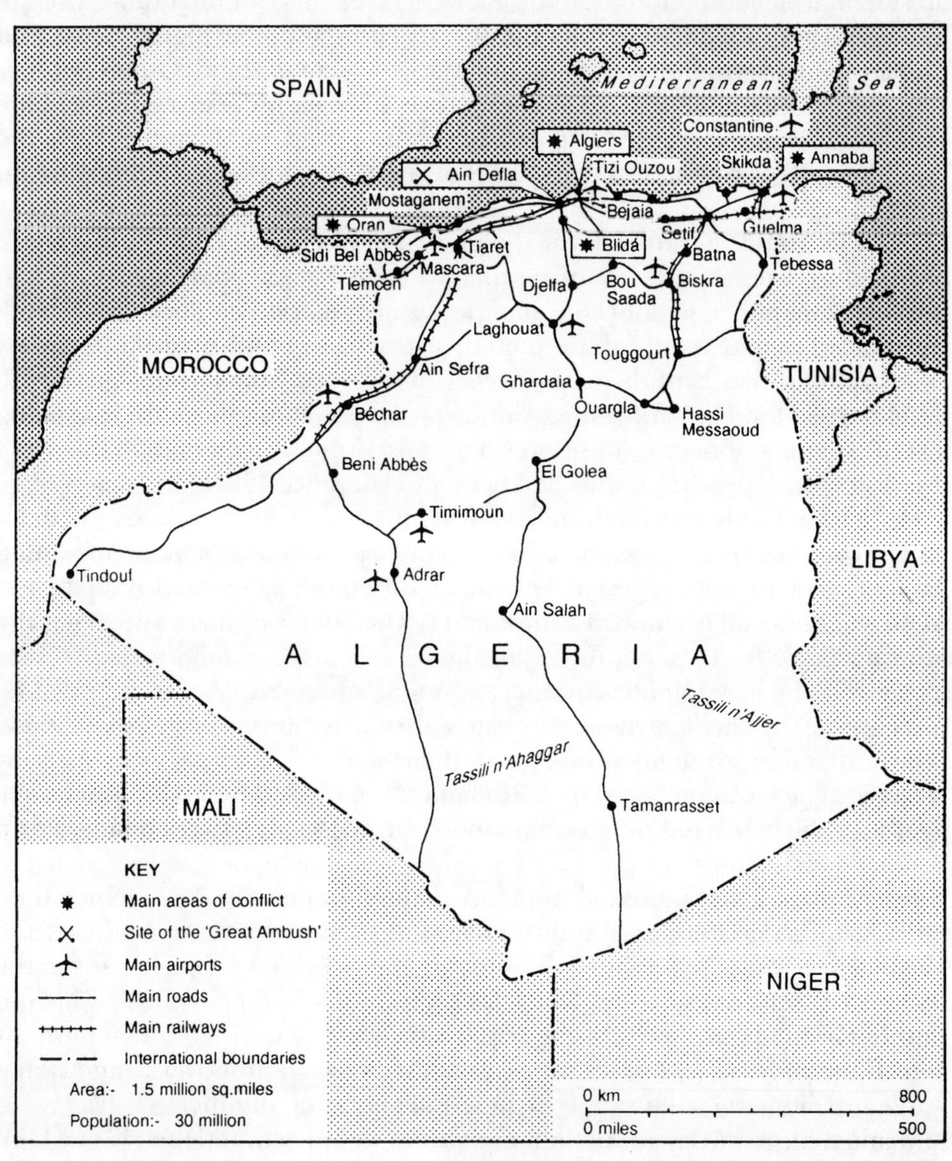

Algeria's Self Destruction

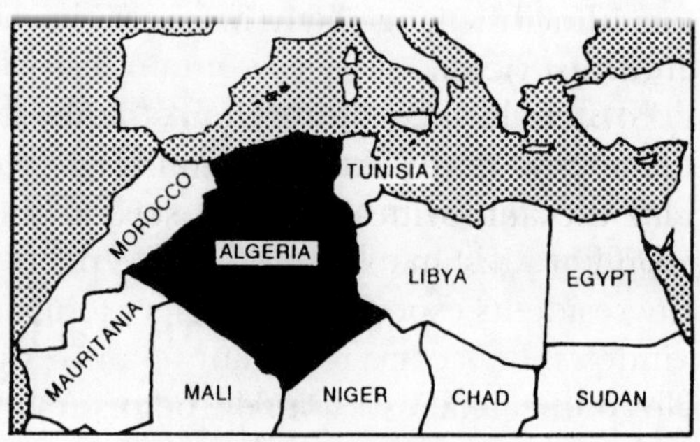

civil war would become more intense. The martyr element made it even more so. Stirred up by their own Algerian mullahs, aided and abetted by Iranian mullahs, every Muslim fanatic is confident of reaching Paradise after ridding the world of an infidel or two. In Algeria, in 1996, anybody who did not support the FIS was an 'infidel'.

Four Boy Scouts were killed and others were maimed by a bomb planted in the cemetery where they were burying one of their scout comrades, killed a few days earlier in another outrage; in the town of Chlef a bomb killed a veteran of the 1954–62 conflict who was attending a commemoration; and in Algiers, Colonel Djelloul Cherif, the city's military commander, was one of 20 people killed during a particularly violent confrontation with militants. The estimate in Algiers was that in the city more than 300 people were dying in political violence each week. 'Bodies aren't being counted any more', an Amnesty International official said.

General Muhammad Lamari, the army's chief-of-staff, said, 'Whatever the price, whatever the sacrifice, the army will wipe out the backward forces of darkness who have betrayed the nation and Islam'. Lamari is one of the leading *éradicateurs* or hand-liners. In response, Anwar Haddam, the leader of the FIS parliamentarians elected in the annulled poll of 1991 and now in exile in the US, said that he and his 'army of idealists and martyrs' would intensify their struggle. They have plenty of arms, ammunition and explosives, supplied by Iran, Libya, Sudan and Saudi Arabia. GIA is also armed with Israeli-made Scorpios and Uzis smuggled into the country.

For ordinary non-partisan Algerians who are sincere Muslims but not fundamentalists the choice is all too stark – an extremist Islamic government or a military dictatorship. The government cannot guarantee the stability needed to attract foreign investment and aid that would boost an ailing economy where unemployment at the end of 1996 was running at 35 per cent. An advisor to the president, Muhammad Merzoug, said, 'It is a real vicious circle. Until the situation is stabilised no one will give us aid and without aid the situation cannot be stabilised.'

In some mosques, pro-government sheiks weep over loud-speakers during Friday sermons as they pray for peace. Policemen disguise themselves as fundamentalists in dangerous attempts to penetrate the fanatics' ranks and sometimes the fanatics themselves impersonate policemen in order to steal cars or 'arrest victims'.

Policemen, like the extremists they are fighting, no longer sleep in their homes and have sent their families to live with friends or relatives. During the year the authorities released several thousand 'ordinary' criminals to make room for Muslim extremists in the prisons. The consequent crime wave forced city residents especially in Algiers to install armoured doors and bars on their windows. Police are reluctant to answer calls for assistance because a frequent ploy of the fanatics is to make bogus emergency calls in order to lead police into ambush.

Intellectuals became another target. Tahar Djaout, a prominent novelist and newspaper editor, was known not to be on the FIS/GIA blacklist but he was shot in the head outside his apartment. A secularist and lover of French literature, Djaout was a 'natural' victim for the extremists, one of whose aims is to stamp out the entire cultural legacy of French colonisation.

The Islamists have appropriated the language of the FLN's war of independence, so that their fighters are Mujahideen or holy war warriors (as in Iran, Afghanistan and Lebanon). Their dead are martyrs and they revile government forces with the term *harkis*, the hateful word for the Algerians who fought with the French army.

Atrocities by Security Forces

The best recruiting agents for the FIS/GIA are the security forces who are guilty of appalling acts of torture. An Algiers university professor was tied to a bench, his nostrils were pinched shut and water was forced through his mouth into his stomach until it was bursting. His army torturers kicked him until he retched, over and over for eight hours. They kept asking him, 'Where are the weapons?' but he had no idea what they were talking about. His case and those of other victims were investigated by French journalists who accepted that they were true. When published, such brutality wins converts to the fundamentalist cause. In the basement of Algiers Chateaueuf police school detainees have their head dunked in sewage and their teeth kicked out. Some prisoners have been burned to death with blowtorches.

'The Islamic fundamentalists are vermin', a senior officer told Italian politicians who attempted to mediate in the conflict. 'We must wipe them out, even if we have to kill millions of people.' It is no more possible to reason with a mind set like this than with the Islamists.

The frontline troops for the government are the *ninjas* – commandos who wear dark hoods to conceal their identity. When security forces, including the *ninjas*, go on patrol they look for young men wearing jeans, leather jackets (which could conceal weapons) and running shoes. Tens of thousands of ordinary Algerian men fit this description so a lot of harmless and innocent men are roughly handled.

The war in Algeria understandably alarms the French government and Spain and Italy are also worried. All three countries fear an Iran-type state at the West's doorstep – 'at our underbelly' as an Italian put it. They predict a radicalised Mahgreb, exporting revolution to Tunisia, Morocco and Egypt. The French, with large numbers of Algerian immigrants, are particularly concerned. A commentator, Jacques de Barrin, wrote, 'From now on the new Algeria war has crossed the Mediterranean'. During 1996, Algerian terrorists planted bombs in the Paris metro, tried to blow up an aircraft over Paris, and committed several murders and acts of sabotage.

The French government is in a difficult position because it cannot condone the Algerian government's excesses and it finds the military's policy of 'eradication' appalling but the terrorists are a greater threat. France is home to 3,000,000 Algerians, including 75,000 professional men and women whose services are badly needed in Algeria.

Paradoxically, this dirty war on Europe's doorstep sank into a black hole of news coverage. It is just too dangerous for journalists to cover. The GIA murders every journalist it can reach. Some, like Tahar Djaout, have been shot in the head in the open, others abducted and never heard of again. The government announced that no further press visas would be issued – an attempt to cover up its own excesses. Almost all overseas organisations withdrew their correspondents.

Djaafer Sei Allah, a leader of the GIA, had promised in 1992 that 'those who fight us with the pen will die by the sword'. He was himself killed by the security forces but other fanatics made good his threat. Arezki Metref, a former editor of the Algerian weekly, *Ruptures*, wrote from exile about the 27 Algerian journalists murdered within 12 months.

> The Islamist assassins make no distinctions. To be a journalist is a crime which merits the death penalty. Those who stay on the job live a half-life in hiding, away from their families. They change their hideouts every evening and their itineraries every morning.

Said Mekbel, editor of the Algiers newspaper *Le Matin*, wrote that a journalist is someone 'who makes a wish not to die with his throat slashed'. The fanatics granted him his wish the very next day – they shot him in the head.

Another 'Colonial Plot'

During 1996 the militants vowed to 'purge Jews, Christians and foreigners, all of whom are part of a colonial plot'. Foreign diplomats live in fortress compounds and in 1996 Britain's six remaining embassy staff could only venture out in an armoured Land-Rover.

Perhaps the most dangerous of the fanatics are the 'Afghans' – the Algerian veterans of the war in Afghanistan who learned all about sabotage and terror from the Mujahideen who brought about the Soviet army's retreat.

In March 1966 armed militants gained access to the Trappist monastery near Medea by claiming that one of them needed medical attention from Brother Luc, an 82-year-old doctor. They then abducted Luc and his six fellow monks. Medea is a GIA stronghold but the abduction was a surprise because the monks maintained friendly relations with the Islamists, calling them 'our mountain brothers'.

The GIA used the hostages to put pressure on the French government to release GIA suspects held in France. On 30 April they announced over *Radio Tunisia*: 'The French broke off the negotiation process so we have slit the throats of the monks. Thus we have executed our threat, as we swore to do before God.'

The whole of France reacted with shock and outrage at the murders and mass protest rallies took place. At Notre Dame Cathedral, Paris, the Cardinal Archbishop extinguished seven candles lit by Christians, Muslims and Jews, saying, 'The candles represented hope for their lives. I wanted them to burn for ever.' The Pope condemned the 'barbaric killings' as 'one of the saddest chapters in the history of Algeria – an offence against God and man'.

Once again the French government called on French citizens in Algeria to leave the country immediately: 'The fate of the monks confirms yet again the extreme danger in Algeria for isolated people, who are particularly exposed.'

On 16 July, the GIA leader, Djamel Zitouni, who had ordered the murder of the monks was himself killed near Medea. It is still not clear whether he was killed as a result of a coup within the GIA, by a rival Islamic group or by the Algerian security forces. Predictably, a GIA communiqué placed the blame on 'enemies of Islam'. The GIA expressed its 'horror' at Zitouni's death. 'He fell into a trap with two other brothers of freedom and all were killed.'

GIA Leadership

Djamel Zitouni, alias Abou Abderrahmane Amine, took control of GIA in October 1994. The son of a chicken salesman from the Algiers suburbs, Zitouni had been one of the thousands of Islamic fundamentalists arrested in 1992. When he was released, the 34-year-old's rhetorical oratory made him a popular leader, especially when he launched a *jihad* (holy war) against Algerian and French targets. He ordered the hijacking of an Air France Airbus on Christmas Eve 1994 and followed it with a bombing campaign in France in 1995. Eight people were killed and 160 injured.

Two days after his death the FIS/GIA 'consultative council' appointed Antar Zouabri, his principal lieutenant, as the new GIA chief. However, Redouane Muhammad Abu Bassir signs many of the terrorist group's communiqués.

It is very likely that a new organisation, the Armed Islamic Group, is not independent but part of the GIA. Algerian military intelligence suspects that it is being used further to complicate the web of fundamentalism in Algeria.

In a population of 28 million, 70 per cent of men aged between 17 and 23 are unemployed and it is from the ranks of these disaffected young men that the terrorist groups find new recruits.

In July the Algerian Defence Ministry appointed Major General Kamel Abderrahmans to be security commander for the oil-rich western region of the country, where the Islamic terrorists sabotage installations, intimidate workers and murder soldiers. A leading figure among the younger generation of officers who had come to prominence after the Islamic insurrection, Abderrahmane was told by President Zeroual to 'waste no sympathy' on the rebels. He probably did not need this instruction: the officer corps, Francophile almost to a man, would allow no FIS man to serve in parliament.

Nevertheless, the government did make concessions in August and September 1996. The FIS agreed to talks with the government at ministerial level and in recognition of this Zeroual ordered the release from prison of Abassi

Madani and Ali Belhaj and put them merely under house arrest. Soon after this it was reported that Madani and Belhaj had begun talks with the government. This report, in the independent daily *El-Watan*, stated that Madani would call for a truce while the government would legalise the existence of the FIS.

These arrangements were not universally welcomed. Redha Malek, a former prime minister, joined other political figures in criticising the release of the FIS leaders as 'a unilateral concession'. In a statement, Malek said, 'The danger is that from concession to concession, the situation becomes worse and finally uncontrollable. Fundamentalists see any concession as a weakness to be exploited.'

Country of a Million Martyrs

In the midst of rising hopes, renewed violence flared. In mid-September 16 civilians were found, beheaded and mutilated, in three areas. Armed men took 12 of the victims from their homes in the town of Sidi Bakhti in the western department of Tiaret and murdered them in a nearby wood. On the same night an armed group slit the throats of three people in Oum el-Bouaghi in eastern Algeria. Another woman, aged 20, was dragged away from her family and was found decapitated not far from her home. Almost certainly, these people were the victims of the Armed Islamic Group.

A constitutional referendum was announced for 28 November 1996. It was on a draft resolution that would outlaw political parties based on religion or language. As the day of the vote drew nearer violence became more extreme and more widespread, though the government-controlled media played it down to support the authorities' claim that they had crushed the Islamic insurgency. Officials were referring to gunbattles and bombings as 'residual terrorism'.

However, on 21 October the mayor of Algiers, Ali Boucetta, was shot dead in the capital, an indication that the rebels were far from crushed. The following day eight people were killed and 30 wounded when Islamists bombed a passenger train west of Algiers. In response, the security forces killed 16 members of the Armed Islamic Group in different parts of Algiers on 28 and 29 October.

On 11 November, 30 well-known figures, including former president Ahmed Ben Bella, made a widely publicised call for peace. The response of the Islamists was to set off a car bomb near a bus stop and school. At least 15 people were killed and 30 wounded. The cycle of violence is endless and some foreign diplomats told me gloomily that they expected the death toll to rise to 100,000 before the end of 1998.

One of the saddest features of the Algerian civil war is that after the end of the war of independence against the French, Algeria was highly respected throughout the Arab world as 'the country of a million martyrs'. Through a failure of internal leadership and external interference, Algeria is now seen as a great social failure. A British diplomat in Algiers said, 'All Algerians share only one thing and that is fear'.

To put this another way, in Algeria there are no moderate or democratic groups or spokesmen within the rebel groups with whom the government could engage in meaningful talks. Similarly, there are few moderates among government ministers and none within the armed forces. Dialogue is blocked by a degree of distrust such as exists in Northern Ireland and Sri Lanka.

The Price of Political 'Victories'

Following scores of atrocities, the government was confident that the national elections called for 6 June 1997 would show that the people could no longer tolerate the Islamist parties. A few days before the elections terrorists exploded two bombs in Algiers, killing 50 people and wounding about 200. The bombs were a 'final warning' to the populace that the opposition parties would ignore an unfavourable result and that that voting was a dangerous business.

Newspaper headlines proclaimed 'Anti-Islamic Forces Sweep Algeria Poll' but the results were not so simple. Of the 380 seats the National Democratic Rally (RND), founded in February 1997 by followers of President Zeroual, took 155 seats; the National Liberation Front (FLN) gained 64 seats. The biggest legal Islamist party, Movement of Society for Peace (MSP) won 69 while the more militant Ennahda claimed 34. Various smaller groupings and independents took the rest. The danger is that while the RND and FLN between them have a governing majority the so-called 'peaceful' Islamists will come under pressure from the militants. In any case, the poll was low, at 65 per cent, because of an FIS boycott of the election. In violent Algiers only 43 per cent of electors voted.

The extremists showed at once that the poll – which 200 international observers said was democratically carried out – meant nothing to them. Mass murders took place in several parts of the country.

The security forces had a much-needed victory on 24 July when Antar Zouabri, leader of the GIA, and considered Algeria's 'most wanted' Islamic extremist, was killed in an army operation west of Algiers. Several of his lieutenants died with him. Apparently, troops had besieged Zouabri in the village of Hattatba. About 100 GIA fighters were killed and 300 surrendered, though these figures, from government sources, were impossible to verify.

Predictably, a GIA spokesman denied that Zouabri had been killed. Muhammad Redouane, known to be a high-ranking GIA killer, called an Algiers radio station to say, 'Zouabri is alive and fighting alongside his troops. The GIA will soon give proof that our leader is alive . . .'

This is a standard tactic of Islamic extremist groups anywhere in the world. By denying the death of a formidable leader they than have time to build up the reputation of a substitute chief. Among extremists, continuity is important. Were the leader to vanish – without a new one to take his place – the morale of the entire group, together with that of its followers, would be badly damaged.

5

Angola's Smouldering War

Following the collapse of Portugal's empire in 1975 a Marxist regime seized power in Angola. Jonas Savimbi, leader of the National Union for the Total Liberation of Angola (UNITA), plotted and fought to bring down the Marxists and a ruinous war continued unabated. Following the end of the Cold War – and therefore of Soviet involvement in Angola – in 1990 the UN was able to bring about elections in 1992. International observers considered them to be fair and reported that the result confirmed the legitimacy of President Eduardo dos Santos and his People's Movement for the Liberation of Angola. Jonas Savimbi disputed the results of the election and fighting continued.

In November 1994 Savimbi and Dos Santos initialled a new peace accord in Lusaka, Zambia. UN negotiators hoped that this would end the bloody civil war in which an estimated one million people had died, with many more maimed – mostly by landmines. Savimbi accepted a vice-presidency for himself and his senior colleagues were given cabinet posts but foreign diplomats feared that Savimbi was merely waiting for an opportunity to go back on his word. The only safeguard was the UN Angolan Verification Mission (UNAVEM), comprising about 6,000 troops from 36 countries.

In January 1996 the UN Secretary General, Boutros Boutros-Ghali, reported that neither the government nor UNITA had begun retiring their fighters, as stipulated in the peace accord. In his report to the Security Council, Boutros-Ghali stated:

> Many of the factors that prevented implementation of the earlier peace accords are still very much in evidence. They include distrust, continuing military activities, foot-dragging over quartering and related activities, obstruction of free movement and the restoration of government administration.

By 'quartering' he was referring to the movement of troops into cantonments were they could be disarmed and then retired from their respective armies into civilian life. Boutros-Ghali said that there had been incidents when repeated failure by both sides to honour their undertakings had cast doubt on the validity of their commitment.

The UNITA command had chosen 15 sites for quartering its troops but only at Vila Nova had it actually placed troops in cantonments, and then only a few. The UN officers on the ground had been expecting at least 150 a day.

Boutros-Ghali, who was voicing the fears of his officers, said that the mistakes which had led to the war's resumption in 1992 were being made again. 'The quartering process must be an uninterrupted and fully verifiable exercise of limited duration', he said. 'It is unrealistic and indeed it is potentially dangerous to keep soldiers in cantonment for a long period of time.'

Various incidents led to increased tension throughout 1996. In May a UN aircraft was approaching the airfield at Andulo, central Angola, where UNITA has a base, when UNITA officers refused permission for it to land. One report says that the pilot was told his aircraft would be fired on. In the dangerous re-routing procedures that followed, the plane and its crew were put at risk. The following month UNITA troops detained a group of five UN peacekeepers at Xinge in Luanda province. The UNITA men, resenting UN instructions, refused to permit the peacekeepers inspection facilities.

Early in July a party of 40 government soldiers attacked an area near a UN demobilisation camp in Benguela province. Several hundred UNITA troops were waiting there to be integrated into the national army. Fighting took place and two government soldiers and some civilians were killed.

The UNITA command claims to have demobilised 50,000 troops and handed in heavy weapons but UN observers have ridiculed this statement. Occasionally co-operation between UNITA and the government improves but for each side this is merely to gain a tactical advantage. Jonas Savimbi is never likely to relinquish his ambition to rule Angola.

Angola's Mercenaries

Angola is hiring mercenaries to provide internal security in return for a share of the country's mineral resources. This information, known for some time to intelligence agencies, was announced officially in a UN report early in November 1996. A UN special investigator, Enrique Bernales Ballesteros, reported that mercenary activities by companies registered as security firms in a third country threatened national sovereignty. He was referring to a company called Executive Outcomes, a firm registered in Britain and South Africa. The UK office is based in Alton, Hampshire.

The government of Angola has a contract with Executive Outcomes to offer protection in return for a stake in the profits from exploitation of the country's natural resources. The mercenaries are recruited in South Africa and Britain. Ironically, Angola has suffered from many attacks by mercenaries since the 1970s. The Ballesteros report states:

> To suggest that some mercenary activities are illegal and others legal is to make a dangerous distinction which could affect international relations of peace and respect between states. Once a greater degree of security has

> been attained the firm apparently begins to exploit the concessions it has received by setting up a number of associates which engage in such activities as air transport, road building and import and export, thereby acquiring a significant if not hegemonic presence in the economic life of the country in which it is operating.

The use of mercenaries by the Angolan government might be unwelcome and dangerous but its wish to do is understandable. Its own armed forces' leadership is corrupt and could not be trusted to maintain security over gold and diamonds. Similarly, it could hardly expect honest security service from Savimbi's men. It seems likely that Dos Santos and his ministers were influenced in their decision to employ Executive Outcomes mercenaries because of that firm's experience in Sierra Leone, where it has 500 mercenaries operating.

War Annuals 1–6 trace the history of the Angolan War, including details of involvement by South Africa, Cuba, the Soviet Union and the United States.

6

Armenia and Nagorno-Karabakh

Armenia was at war with Azerbaijan for eight years, 1986–94, over the disputed territory of Nagorno-Karabakh. Even during the disastrous earthquake of 1988 and the country's economic collapse after the break-up of the Soviet Union in 1991, the Christian-Armenians versus Islamic-Azeria conflict continued but it had deeper roots. In 1915 the Turks, also Muslims, massacred 1.5 million Armenians and their descendants, naturally enough, felt that they could never again trust Muslims.

Under Soviet control, Nagorno-Karabakh was a mainly Armenian-populated enclave in Azerbaijan. Following the truce of 1994 all the Azeris departed, leaving a population of 130,000 Armenians. Should Azerbaijan ever seek to reimpose its sovereignty over Nagorno-Karabakh a slaughter would result.

The Armenians have turned the former enclave into a form of fortress and have anticipated Azeri aggression by building a strategic military road through Lachin, in the centre of the narrow waist of a region which has a mixed Azeri-Armenian population. They have also heavily resettled the villages along the Lachin road. Other new roads have been built and old ones resurfaced.

A military-minded people, the Armenians, especially those in the enclave, are plentifully supplied with weapons, mainly of Russian-make. Armenians abroad, notably those in the United States and France, have funded the country's defence requirements, which include surface-to-surface missiles capable of hitting Baku, the capital of Azerbaijan, and its vast oil installations. The Azeris are well aware of this but they continue to press for the return of the towns of Agdam, Fizuli and Jebrail. A tentative proposal put forward by Armenia is that the Azeris could have a narrow corridor along the Iranian border to link their own enclave, Nakhichevan, with Azerbaijan proper. The price would be Armenian ownership of the gold-mining town of Kelbajar.

Armenia is not at peace with itself and violence erupted in the capital, Yerevan, late in September 1996. It followed the re-election, on 22 September, of President Ter-Petrossian, who apparently won 52.9 per cent of the vote against the 41.7 per cent of Vazgen Manukian, the opposition leader. Demonstrators claimed that the elections were rigged and stormed the parliament building, where they beat seven deputies, including the chairman of the

Nagorno-Karabakh

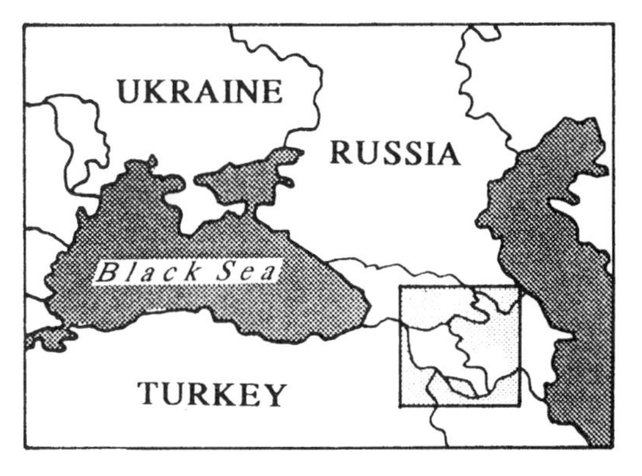

parliament. Another 50,000 people demonstrated in the city centre. In a swift reaction, Ter-Petrossian banned unauthorised demonstrations and marches and removed the immunity from arrest from eight opposition deputies who had taken part in the demonstration.

Soldiers and tanks were on the streets and security was strict around the television centre, the target of previous unrest. Certainly Armenia needs a strong president and especially one with a firm and workable foreign policy. Turkey, in the west, is still hostile, as is Azerbaijan, to the east. The Georgian government is friendly but is crippled by a war in Abkhazia, through which Armenia's rail link with Russia passes. Nothing can travel along this line.

Fundamentalist Iran is irritated that Armenia receives American aid worth £50 million a year but nevertheless trades briskly with Armenia. President Ter-Petrossian regards the Iranian regime as a menace and refuses to meet its leaders. It is well known that Iran would like to see the Azeris return to Armenia.

The key to Armenia's security remains with Russia, which has 15,000 troops in Armenia and helps to guard its borders. This is not an entirely altruistic policy. Russia needs to protect its rear in the Caucasus and it does so partly by shielding Armenia from aggression.

7

Bangladesh War of Genocide

40-YEAR CONFLICT

Background Summary

A mistaken policy of the British Government began the conflict which was to become a genocidal war. When India and Pakistan became independent from Britain in 1947, the Buddhist tribes of the Chittagong Hill Tracts expected their land to be recognised as a native state or as part of a confederation with tribal areas of north-east India. But the British foolishly incorporated the Hill Tracts into Pakistan, a Muslim state. The 600,000 Buddhists of the Tracts, a total of 32 tribes, feared for their future under expansionist, racist Islam.

In fact, they came under no great pressure from Pakistan but everything changed in 1971 when East Pakistan and West Pakistan fought a bloody civil war. The major consequence was that East Pakistan became Bangladesh and the new government began to drive out the Buddhist tribes in order to give their land to Muslim settlers from Bengal. The Tracts make up about ten per cent of the area of Bangladesh.

The Buddhists, normally a peaceful people, formed the *Jana Sanghati Samity* (JSS) as a 'self-defence association'. The Chakmas, the largest tribe, created a military wing, the Shanti Bahini. Unexpectedly, the Shanti Bahini held out against the Bangladeshi army and in 1984 Major General Noor Uddin Khan, commanding the 24th Division, the so-called Bengal Tigers, was ordered to find 'a permanent solution to the Hill Tracts problem'.

His solution was the Nazi-type one of terrorism, deportation and genocide. When the army's cruel methods were still not effective enough the government brought in vicious Muslim zealots as mercenaries to hunt the Shanti Bahini. The ploy failed; the guerrillas killed 300 of these men in 1990.

The downfall of President Hussein Ershad in December 1990 and his replacement by a woman ruler was expected to produce more humane treatment for the Buddhists but this did not happen. The Shanti Bahini had received arms and ammunition from India and China but these sources dried up by 1991 and they were forced to rely on what they could procure by raids on army bases and by ambushing patrols.

Various ceasefires were declared but they were invariably tactical ruses by the government to gain some advantage. The most significant ceasefire was agreed in August 1992. The tribes, now amalgamated into the *Parbatta Chattagram Jana Sanghati Samity* (PCJSS) or Chattagong Tracts Peoples' Solidarity Organisation, continued to press for autonomy in their ever-shrinking area but only the Chakmas still sought a solution by armed struggle.

One consequence of the unrest, especially following Shanti Bahini attacks against Bengali settlements in 1986, was the creation of a Chakma refugee problem in the neighbouring Indian state of Tripura. Shanti Bahini fighters hide among this exiled population and they receive clandestine support from India. About 50,000 Chakma refugees live in six camps. Bangladesh and India agreed on a repatriation programme in 1993 but it has moved only slowly.

A more encouraging development is that the dreaded 24th Division is heavily involved in development work, including the building of schools, roads and general infrastructure. This change of policy – though human rights abuses continue – gave some impetus to peace talks and ceasefires were extended from time to time. Peace talks were then delayed by two years of political bickering among Bangladesh's political parties. This ended only in June 1996 when the Awami League won the elections, bringing to power a government led by a woman, Sheikha Hasina.

Before the new government could act on the Hill Tracts problem an incident occurred that put the entire peace process in jeopardy. On 10 September Shanti Bahini rebels massacred 35 Bengali loggers in an isolated part of the Rangamati district. Two wounded loggers escaped to raise the alarm. Bangladeshi troops, supported by helicopters, at once began a search for the guerrillas and for three missing loggers.

The attack was aimed at putting pressure on the Hasini government to address the Buddhists' long-standing grievances. It certainly brought attention to the area but not in the way that the Shanti Bahini intended. Those tribes more peaceful than the Chakmas and ready to compromise with the Bengali settlers were profoundly angry with the Shanti Bahini because they feared a backlash from the Bengalis. Indeed this is what happened and to such an extent that the army was forced to intervene between the two sides.

There seems to be no end in sight to the conflict, especially as now Thailand is smuggling arms by sea into the Hill Tracts for use by the Shanti Bahini.

8

Bougainville's War of Independence

BREAKAWAY ISLAND

The mountainous and jungle-clad island of Bougainville off the eastern end of the great island of Papua-New Guinea (PNG) lends itself to guerrilla activity. During the Second World War the Japanese army of occupation failed to capture Australian coastwatchers and the invaders suffered heavy casualties at the hands of Australian and native guerrillas.

In 1989 the 170,000 inhabitants of Bougainville and the smaller Buka Island sought independence from the PNG government, which refused to consider the secessionist request. A bloody uprising by the Bougainville Revolutionary Army (BRA) in March 1990 forced the PNG government to withdraw its troops from the islands but they were sent back when the rebels set up their own administration. The PNG government had no intention of giving up the wealth created by Bougainville's copper mines.

Unable to defeat the BRA militarily, the government embarked on a blockade of the island, leading to great hardships among the people. It has the means to enforce a blockade – four Tarangau-class patrol and coastal boats and two amphibious craft, Salamaua-class, all built in Australia. Australia also supplied four helicopters, which the PNG Defence Force turned into gunships and used in the 'Saint Valentine's Day massacre', in February 1990.

The government incites existing divisions among Bougainville's groups. The people of south Bougainville see their area as a southern state within an independent Bougainville. Like the people in the BRA stronghold of central Bougainville, the southerners insist that any agreement must include the total and permanent withdrawal of the PNG army, but they are less hardline than the central populace in their approach to negotiations with the government.

Early in 1994 the PNG Defence Force won back the key Panguna copper mine and the area around it, but this did not constitute a decisive victory.

The War in 1996–97

An 18-month ceasefire, supposedly to allow time for peace negotiations, ended in March 1996 and tensions immediately began to rise. In June PNG raiding

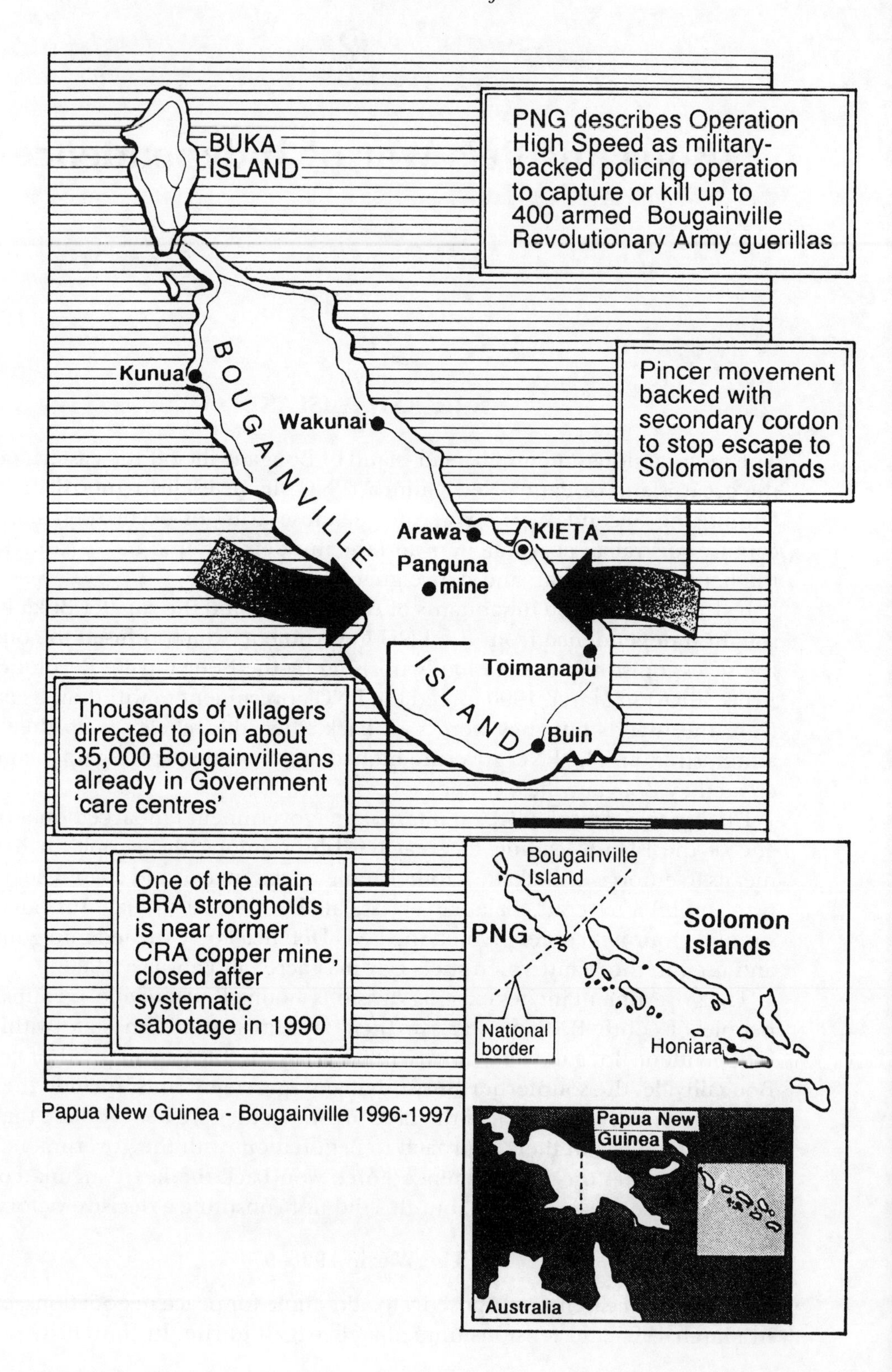

Papua New Guinea - Bougainville 1996-1997

parties attacked BRA refuges in the neighbouring Solomon Islands. These raids were a precursor to 'Operation High Speed', the biggest government push of the war. About 1,500 men were involved, principally from the two battalions of the Royal Pacific Islands Regiment.

Through its foreign affairs minister, Alexander Downer, the Australian government of John Howard expressed its alarm and concern and called the offensive a 'disturbing development'. The PNG government of Sir Julius Chan responded angrily to this 'unwarranted interference'. Even so, he accepted £100,000 of 'non-lethal' equipment for his Defence Force.

Just before the offensive the government tried to induce about 25,000 people living in the probable combat area to join about 50,000 others resident in 'care and safety centres' run by the government. The plan met with only limited success but the government went ahead with the campaign, one objective of which was to block BRA escape routes to the Solomons.

Far from being alarmed by the coming offensive the BRA responded aggressively. The BRA commander, 38-year-old Sam Kauona, issued a media statement: 'The BRA welcome the declaration of all-out war and is waiting to meet the PNG forces.'

Operation High Speed began on 21 June 1996 and within a month the BRA claimed that it had driven the PNG Defence Force from its beachhead and blocked the offensive. The government explained, unconvincingly, that its forces merely made a 'phased withdrawal' in preparation for further attacks. The simple fact is that the BRA had been victorious. In September BRA raiders massacred 12 soldiers at Kangu Beach in Southern Bougainville.

The army's morale was severely damaged, so much so that some soldiers refused to return to active duty in Bougainville. Brigadier Jerry Singirok, Chief of the PNG army, dismissed 19 of them and announced that this was only a small step in a major plan to raise the standards of discipline, training, management and morale. A spokesman for the soldiers, quoted in PNG's daily newspaper was bitterly critical of the army's leadership and the government's 'neglect' of the army:

> We are being sent to fight a war without food, boots and socks, clothing and adequate weapons. In a company of 120 men we should have six or seven specialist weapons, such as machine-guns. In fact, we have only two or three operable guns. Every soldier should have at least two grenades when in action but we have none.[1]

Brigadier Singirok admitted to journalists that the army was having logistical problems but that the situation would improve in 1997. He said: 'The fact that we are able to sustain operations in Bougainville even up to November [1996] is a big achievement and we hope to get better at it'.[2]

The doubt implicit in this statement could hardly have cheered his men but the government's announcement that it had bought two commercial Bell 101 helicopters for transforming into gunships may have encouraged them. The army had been using old and often unreliable Iroquois helicopters, a gift from Australia.

The poor morale of the PNG Defence Force and its apparent lack of professionalism in the field surprises the Australian army, which has provided 'advisors'. At one time 110 Australian instructors were working with the PNG army. I have been told by an Australian officer:

> These men just do not make good soldiers; they have neither the aptitude or the attitude necessary. There is an engineer battalion but it is difficult to teach them the basic skills.

The Australians are forbidden to become directly involved in advising the PNG Defence Force on tactics to counter the BRA insurgency. An official source admits that one senior Australian officer was recalled from PNG by the Labour Government [in power until March 1996] for disobeying this instruction.

The conflict in Bougainville is a sniper's war – 'a dirty sniper's war' according to the PNG Prime Minister, Sir Julius Chan. Most of the PNG army's 200 deaths have been the result of a single shot from an unseen attacker, often when the victim was not in combat. Snipers of the 2,000 strong BRA seek out enemy troops washing clothes by a river bank or when they are using a latrine.

Politics of the Dispute

Before PNG became independent of Australian administration in 1975 the Australian government gave approval for a giant opencut copper mine to be developed at Panguna. The mine provided a large part of the new government's income and foreign earnings. Bougainville had its own provincial government and development corporation but Panguna clans protested about the 'unfairness' of the distribution of mineral royalties as well as land and environmental damage. By 1990 these problems had broadened to a demand for total independence. A former officer of the PNG Defence Force, Sam Kauona, who had been trained in the use of explosives in Australia, blew up power pylons at the mine, which was closed because of this and other sabotage. There followed army raids, rebel ambushes, dreadful human rights violations, abortive and botched military offensives and equally futile peace talks.

PNG Prime Minister Rabbie Namalui, influenced by the military, applied a blockade which had no effect whatever on the guerrillas but, because it kept out medical supplies, caused the deaths of hundreds of civilians. Piais Wingti, who succeeded Namalui, lost the goodwill of even the reasonable men among the Bougainville secessionists. Sir Julius Chan, who in 1994 became prime minister for the second time, met Sam Kaounda in the neutral Solomon Islands for peace talks. For a time these seemed promising, but a BRA man, on his way to the talks, was killed by security forces and as a result the entire BRA conference contingent stayed away. In a desperate effort to sustain the conference, Australian Prime Minister Paul Keating met the expenses of a regional security force to protect the venue but trust between the parties – the PNG government and the Bougainvilleans – no longer existed. In January 1995 Pope John Paul, on a visit to PNG, publicly called on Bougainvilleans to 'remove bitterness from their hearts' and to lay down their arms. Unsurprisingly, he was ignored.

The Australian government, late in 1995, sponsored peace talks in the northern city of Cairns and provided the BRA delegates with safe passage. The meetings seemed hopeful and after the December talks the BRA leaders stayed for a time in Honiara, Solomon Islands, for a holiday, before taking a circuitous route home to Bougainville's south-east coast. PNG troops, in direct breach of ceasefire terms, blocked their arrival and attempted to search their boat for weapons. There was a fire-fight but nobody was injured. However, once again any trust that remained was destroyed, as Sir Julius Chan effectively demonstrated in March 1996 when he revoked the ceasefire.

On 12 October the prime minister of Bougainville, Theodore Miriung, was assassinated and in November Gerard Sinato was elected to replace him. He was taking on a confused political situation, with the Bougainville Interim Government and the Bougainville Transitional Government competing for power. The Transitional Government has members from all over Bougainville while the Interim Government is the political wing of the Bougainville Revolutionary Army. A natural peace-maker, Sinato quickly engaged in talks with the Port Moresby government.

As the most powerful nation in the region, Australia's inability to stop the Bougainville war is probably its most serious foreign policy failure since the country's involvement in the Vietnam War. Since 1990 Amnesty International has been documenting violations of human rights by government troops and the BRA guerrillas. They include 'executions' after capture, torture and rape.

There can be no military solution to the Bougainville problem, if only because neither side can win. The BRA has neither the numbers nor the equipment to wage a pitched battle against the PNG Defence Force or to prevent the enemy soldiers from landing. But from the mountainous jungles the BRA, with popular support in many villages, could continue to pick off the PNG troops for years to come. Almost inevitably, Bougainville will gain its independence, and equally inevitably many more people will die before this happens. Bougainville, like much of the south Pacific, has a 'payback' culture, similar to the vendettas of Latin countries. A killing must be avenged, a crime atoned for. This gives added momentum to the increasingly bitter war.

References

1 *The National*, published in Port Moresby, 18 November 1996.

2 Quoted by Mary-Louise O'Callaghan, South Pacific correspondent for *The Australian*, 22 November 1996.

9

Burma (Myanmar): The Karens Under Pressure

The struggle by Burmese minority groups, notably the Karens, has been well documented in earlier editions of *War Annual.* On several occasions during 1995 and 1996, foreign observers and even diplomats in Rangoon wrote them off as 'totally defeated'. They were probably unduly influenced by propaganda emanating from the State Law and Order Restoration Council (SLORC) which wanted to give the impression that peace had descended on Burma after fighting that had gone on since 1949.

In 1995 the SLORC junta brought off a damaging coup against the Christian-led force of General Bo Mya by inducing Buddhist Karen to defect from his leadership. With the backing of SLORC, the Buddhists formed the Democratic Karen Buddhist Army (DKBA). The use of the word 'democratic' was a smokescreen since democracy had nothing to do with the DKBA's policy or practice. On the SLORC's urging, the DKBA launched terror attacks on the refugee camps on the Thai border after months of warnings to the camp-dwellers to return to Burma or see their camps destroyed. On the night of 29–30 January, the Buddhist terrorists torched three camps inside Thai territory. More than 4,500 people were left homeless.

The DKBA also partly destroyed even larger refugee camps in the border town of Mae Sot. The SLORC told the Thai authorities, through its defence attaché in Bangkok, that attacks would be made. This had the desired effect of keeping the Thai military on the sidelines. It was obvious to observers in Rangoon and Bangkok that the Thai government hoped that the problem would be resolved by 'natural means' – as one diplomat put it. He meant that the Burmese refugees would return home under duress but apparently voluntarily. In February 1997 the Karens, under intense military pressure from the army, burnt and abandoned some of their bases rather than allowing them to fall into the SLORC's hands.

The base of the Karen National Union (KNU) at Htikapier came under attack by the Burmese army in mid-February 1997 and reports from neighbouring Thailand stated that the base had been overrun. However, this was only a temporary setback because the Karens, from support bases in Tak Province, 260 miles north-west of Bangkok, regrouped and drove the soldiers out of Htikapier.

The KNU was the only major ethnic insurgency to have reached an accord with the SLORC but after the attack on their base the accord fell apart. The Thais were alarmed about escalating war which, a military spokesman said, would cause an influx of 20,000 more refugees into Thailand. At the beginning of 1997 more than 80,000 Karens, most of them KNU supporters, had already found refuge in Thailand.

Meanwhile tension was increasing elsewhere in Burma. An Amnesty International Report on 14 February 1997 stated that 1996 had been the worst year of the decade for violations of human rights. More than 2,000 human rights protesters had been arrested and severe restrictions had been imposed on the movements and speech of the popular political leader, Ms Aung San Suu Kyi. According to Amnesty, attacks on members of Suu Kyi's National League for Democracy (NLD) had been organised by SLORC.

'Throughout 1996 and in 1997', Amnesty said, 'SLORC suppressed peaceful political meetings, gatherings and demonstrations. At the beginning of the year, Burmese people could gather peacefully every weekend to listen to speeches by NLD leaders but now even this has been forbidden.' Amnesty officials said that they were particularly worried about violent physical attacks on senior members of the NLD. 'On 9 November about 200 young men attacked Aung San Suu Kyi's motorcade with iron bars and sticks. They were thought to be members of the Union Solidarity Movement (USDA), a SLORC-sponsored group.' Shortly before the attack on her motorcade, Ms Sue Kyi had made public the information that a government minister had told a meeting of USDA members that she should be killed.

At a meeting of seven south-east Asian countries, together with 15 European nations, on 14 February, Burma, which wants to become part of ASEAN, was strongly criticised. Malcolm Rifkind, the British Foreign Secretary at the time, spoke about the 'disturbing situation' in Burma. The European nations have long urged greater pressure against the military regime while ASEAN pursues a policy of 'constructive engagement'.

The SLORC survives because it has a large and well-trained army, concentration camps for its political prisoners and the most oppressive regime in south-east Asia. However, any attempt to kill Suu Kyi would bring about a state of civil war that even the SLORC could not control.

10

Cambodia's Torments

Background Summary

The ghastly conflict in Cambodia has been well documented in all editions of *War Annual* but it needs to be stated again that the country's modern problems began with the withdrawal of the French from Indo-China, following the ignominious defeat of the French army at Dien Bien Phu in 1954. The Vietnamese saw themselves as the natural successors to control this great and complex region. The Americans, seized by paranoia about 'the Communist threat', attempted to keep the North Vietnamese Communist regime out of South Vietnam, Laos and Cambodia, but the superpower was militarily defeated and humiliated in the Vietnam War, 1959–75.

In Cambodia, the Khmer Rouge Communists under Pol Pot defeated the US-supported government and in December 1978 the all-conquering Vietnamese sent their armies into Cambodia to protect 'the south-western flank from "foreign dominance"', meaning China.

The Soviet Union, the enemy of China, backed Vietnam, while China, needing an ally in Cambodia, took Pol Pot's Khmer Rouge under its wing. This infamous leader butchered two million people – out of a total population of seven million. The victims were those who opposed – or were suspected of opposing – Pol Pot. Tens of thousands of children were murdered because Pol Pot feared that they might, at some time in the future, become his enemies.

President Heng Samrin put Cambodia's army under Vietnamese command, the better to oppose the Khmer Rouge, which, with two opposition non-Communist groups, took to the jungle. These were Son Sann's Khmer People's National Liberation Front (KPLNF) and Prince Sihanouk's *Armée Nationale Sikhanoukienne* (ANS).

After the Vietnam army withdrew in 1990 the Khmer Rouge began a 'People's War', actually a continuation of Pol Pot's campaign of genocide. The United Nations Transitional Authority in Cambodia (UNTAC) became operational on 31 October 1991 and, under Lieutenant General John Sanderson of Australia, this peacekeeping force achieved considerable success before its role ended in mid-1993.

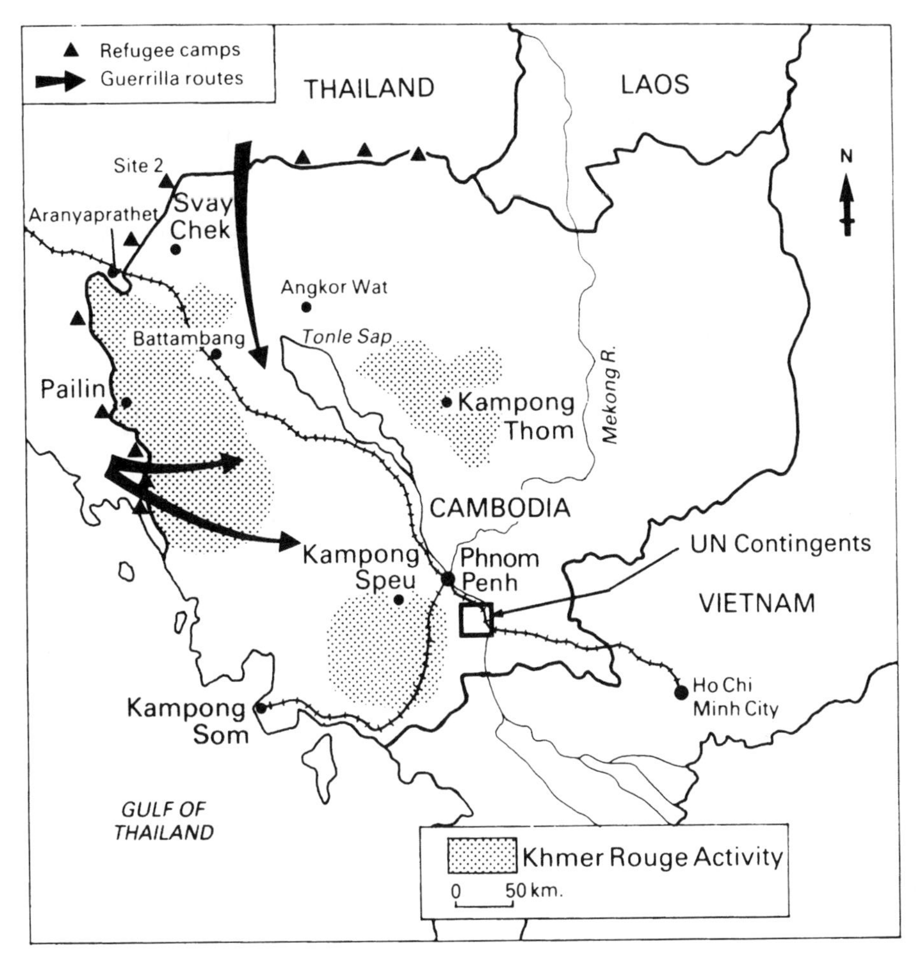

Cambodia: Khmer Rouge Activity Continues Despite Peace Efforts

Throughout 1994 and 1995 the Khmer Rouge remained undefeated in the field, despite efforts by several countries to make the Cambodian army more professional. For instance, Indonesian Special Forces trained 225 Cambodian paratroopers. North Korea offered massive military help but for fear of offending Western aid donors the Cambodian government rejected the North Korean deal.

The War in 1996–97

One of the most significant events in Cambodia's long struggle against the terrorist armies occurred in July 1996 when a jungle warfare training school began operations. Australia paid the equivalent of US $1.2 million to build the school, the Sala Aranya Songkriam, including 14 buildings for barracks, lecture rooms and a mess hall, together with a water treatment plant. Australia also met the cost, US $1.3 million, of training 55 Cambodian instructors. After weapons, the training of instructors is always the most expensive part of military aid.

The curriculum is similar to that of the Australian Defence Force Land Command Battle School at Tully, North Queensland, where the Cambodians were trained. In the Cambodian army a battalion numbers about 250 officers and men and, in rotation, entire battalions spend a month studying how to use and maintain weapons. There is also instruction in 'guerrilla psychology', which is designed to help the soldiers anticipate the actions of the Khmer Rouge fighters.

A second month's training is spent at jungle camps close to the training centre and here the battalions learn tactics, camouflage, demolitions and booby traps. Working as platoons, the trainees spend a final month on realistic manoeuvres.[1]

Cambodia has Co-Defence Ministers, Generals Tea Banh and Tea Chamrath, and both spoke at the inauguration of the new Jungle Warfare Training school, which is situated at Pich Nil, south-west of Phnom Penh. Tea Chamrath announced, perhaps too ambitiously, that the new training would bring about the end of the insurgency forces.

This could hardly happen in the short term since the Royal Cambodian Armed Forces (RCAF) needed to train its three Counter-Guerrilla Brigades, each of 1,000 men, in the new techniques. However, this was the first attempt to bring real professionalism to the RCAF. Weapons at the Sala Aranya Songkriam have come from a number of sources, including Israel and Brazil.

About the time of the publicity given to the jungle warfare training school, Khmer Rouge guerrillas were seizing hostages from villages in the southern province of Kampot. One group was taken during an attack on a sawmill. Those villagers left behind were told that their friends were being 'sent to be re-educated to clear their minds'. It was not clear whether rebels were demanding a ransom for these and other hostages or whether they were being used.[2]

On 8 August it became evident for the first time that a split was taking place among the senior leaders of the Khmer Rouge. The initial indication of this important event came from the Khmer Rouge's radio station when an unnamed spokesman read a fierce denunciation of Ieng Sary, long regarded by many Cambodian government officials as Pol Pot's principal deputy. Within hours the nation's Second Prime Minister, Hun Sen, announced that two senior field generals, their troops and their supporters had defected to the government. They were Major General Sok Pheap and Mit Chhean. Pheap is chief of 'Division 450', whose headquarters were at Phnom Malai on the Thai border, from where he controlled much of the western Cambodian region. Chhean is commander of 'Division 415' and ruler of the city of Pailin, a key factor in the Khmer Rouge's economy; it is the centre of timber and mining industries.

According to Cambodian army intelligence, Pheap and Chhean between them control 3,500 guerrillas and 35,000 'others', probably families and civilian workers. These figures could well have been inflated so that the government can claim even greater prestige from the defection.[3]

In his broadcast, Hun Sen – himself a former Khmer Rouge regimental commander until his defection in 1978 – said that months of negotiations had preceded the defection. General Sok Pheap almost at once called a press conference – a rare event for a Khmer Rouge leader – to explain the situation. He said: 'Forces under Ieng Sary have split from the Khmer Rouge's extremists but we are some way from a peace deal.'

The government's position seemed to be that the forces of Pheap and Chhean would remain in place at Phnom Malai and Pailin, 'under observation' from Cambodian army officers. There was also an intriguing reference to another Khmer Rouge formation, 'Division 519', led by Ta Sue, which was expected to break away from the guerrilla movement.

Leaving the defecting forces in place in the key areas that they had long controlled made sense. If they were to move out a vacuum would occur and hardline elements of the Khmer Rouge would move in. As the news of the defection broke much attention was being given to the position of Son Sen, the Khmer Rouge 'defence minister'. At one time he was believed to have had secret talks with the Phnom Penh government with the object of negotiating peace. Foreign diplomats in Phnom Penh reported that various government ministers were 'working assiduously' to widen the rift within the Khmer Rouge's ranks.[4]

Pol Pot was 'definitely alive' at this time, according to Thai sources, who said that he was operating from Anlong Veng, in northern Cambodia. This is the base of General Ta Mok, probably the most extreme of all Khmer Rouge chieftains after Pol Pot himself.

The apparent break-up of the Khmer Rouge is not caused by an ideological split but by disputes over money and property. Ieng Sary is supposed to have plundered the fortune of the movement and to have shared the takings with Sok Pheap and Mit Chhean. In any case, the areas controlled by these three is much more productive and profitable in gems than the Anlong Veng region. Overall, the Khmer Rouge is estimated to earn £6.5 million each month from the sale of rubies, sapphires and hardwoods. This is why the warlords can afford to buy weapons and many luxuries.

Thailand is deeply interested in the peace negotiations between the government and the Khmer Rouge breakaway factions. If the Phnom Penh government recognises the breakaway leaders and thus confers legitimacy on them, Thai businessmen will once more be able to trade openly in Cambodia's natural resources, commerce that was broken off after the Khmer Rouge withdrew from the UN peace process in 1993.

A UN official in Cambodia says that the Khmer Rouge will be even more dangerous to the government if it joins the political process leading to elections in 1998. A Khmer Rouge alliance with either of the two coalition parties would give that party a much greater chance of electoral success.

Following the Khmer Rouge internal split, the First Prime Minister, Prince Norodom Ranaridd, said that he would want to repeal a ban on the Khmer Rouge imposed in 1994; it would then be possible, he said, to approach the problem of national reconciliation.

Killers as 'Heroes'

The difficulty is that Ranaridd and Hun Sen, the Second Prime Minister, are bitter enemies and their animosity has paralysed concerted action against the Khmer Rouge. The entire government is feud-ridden and this prevents firm decision-taking. It will come as a surprise to people who have suffered at the hands of the Khmer Rouge – such as the families and friends of murdered Western hostages – that some Khmer Rouge killers are now fêted by the government. One of them, Chhuk Rin, who killed 13 passengers in an attack on a train and took three Western hostages, is a 'hero'. This is because Rin, aged 45, defected from the Khmer Rouge. Certainly he went over in the midst of the hostage crisis and before the hostages were killed but the government did more than merely accept him. Rin was made a lieutenant colonel in the Royal Cambodian Armed Forces, given a 'gratuity' of several thousand dollars as well as a car and a luxury home. A bodyguard of 20 came with the material rewards. Interviewed in August 1996, Rin said:

> The government will not defeat the Khmer Rouge in a guerrilla war. They will defeat them by developing rural villages like mine – so the Khmer Rouge men will say, 'Oh, look at that village! Why should we stay in the forests and fight?'

He conceded that the Khmer Rouge abused the human rights of the Cambodians and of Westerners but he blamed the governments of Britain, Australia and France for failing to co-operate effectively with the Cambodian government to bring about a deal for the hostages' freedom. Also, he said, the hostages' families could have done more to put pressure on their governments.[5]

Quite what the families could have done in addition to their pressure and pleas is difficult to see. Chhuk Rin was obviously trying to find anybody to blame other than himself. His transition from cruel bandit leader to national hero is one of the more sickening aspects of the Cambodian civil war.

In mid-September 1996 Ieng Sary emerged from the jungle to ask for an amnesty in return for defecting to the government with about 4,000 supporters. Laughing happily for the press photographers, Sary said:

> I have nothing to be sorry for because I had nothing to do with ordering the execution of anybody or even of suggesting such a thing. I even saved some people who might otherwise have been executed.

Cambodia's two prime ministers conferred and then recommended to the king that Sary be granted an amnesty, which he was.[6] Nobody believes Sary's protestations of innocence but the government was finally ready to do a deal with any Khmer Rouge member for the sake of dismembering and finally dismantling the organisation. Co-Prime Minister Norodom Ranaridd said:

> Ieng Sary does not have to become a normal citizen, he only has to keep a very low profile. He will certainly not become a member of the government.[7]

Many foreign diplomats are sceptical about this assurance. One said, 'An amnesty for former genocidal war criminals is a strange idea. If this can happen it is all too possible that Ieng Sary and other Khmer Rouge leaders could enter the government or at least be employed by the government.'[8]

The International Cambodian Genocide Programme (ICGP) research team documenting the atrocities committed by the Khmer Rouge points out that in Bosnia a few score mass graves have been found and that the world is outraged as a result. In Cambodia tens of thousands of mass graves have been located and mapped without an equivalent outcry from the civilised world.

In October 1996 two more Khmer Rouge bases were reported to have fallen. General Sou Kimsun, a spokesman for the army, said that about 2,000 fighters and 10,000 Khmer Rouge civilians had surrendered to the government after being encircled for five days. Unfortunately, the government's publicity machine is so intent on claiming victories that its figures are always suspect.

Cambodian society, still traumatised by Pol Pot's genocide, faces new tensions. Rank and file Khmer Rouge defectors are now living among the very people whom they victimised and tortured. Far from feeling any remorse, some of these killers are making money from their notoriety. A former guard at the infamous Toul Sleng torture chamber charges foreign journalists $100 for an interview.

Meanwhile, the Khmer Rouge die-hards, operating from their Thai border jungle bases, continued to raid villages and attack army patrols and isolated bases. Even if Pol Pot were proved to be dead it was unlikely that his terror organisation would die with him.

Pol Pot Dead – Again

Early in June 1996 came yet another report, perhaps the tenth, that the monster of Cambodia was dead. The report coincided with the end of the ICGP which had conducted a thorough scrutiny into Pol Pot's activities financed by the US and Cambodian governments. It was possible that Pol Pot and his associates believed that his 'death' might reduce the impact of the Programme's report.

The Times of London ran an interesting editorial on 'Cambodia's Mass Killer' on 8 June 1996. The leader-writer reported that since the investigation had started more than a year earlier it had located thousands of mass graves and had interviewed witnesses who were only now summoning up the courage to tell their stories.

> The Khmer Rouge was almost as meticulous as the Nazis in keeping detailed records of massacres, often matched by photographs. It [The ICGP] now believes that the secretive Organisation on High led by Pol Pot starved to death or butchered by the most brutal methods at least two million Cambodians, double previous Western estimates.

The Times article criticised the vote at the UN which recognised a coalition, of which Khmer Rouge was a part, as the legal representative of 'Democratic Kampuchea'. In the writer's estimate, 'the final peace settlement held the door open to the Khmer Rouge, which then kicked the door in'.

The Times, from its many sources of information, concluded in its hard-hitting editorial that:

> The remnant of the once-strong guerrilla force is still capable of keeping alive a doctrine of terror which emptied Cambodia's cities and turned its fields into a vast concentration camp. The intellectual authors of this sustained atrocity, including Pol Pot's close associates Ieng Sary and the 'one-legged butcher' Ta Mok, are still at large on the Cambodian–Thai border.

The essence of *The Times* editorial argument was that the evil men had to be tried and convicted before they died. 'From the grave, Pol Pot's smile will haunt Cambodia still. In court, his mystique might at last be dispelled – because then there could be no second coming. There are monsters who are best taken alive.'

Pol Pot was *not* dead. In June he was supposed to be in the process of surrendering his remaining rebel troops, but he changed his mind. He then ordered the execution of the former Khmer Rouge defence minister Son Sen and his family for alleged 'treason'. According to Cambodia's first Prime Minister, Prince Norodom Ranariddh, 11 people in all had been killed and Pol Pot then drove over their bodies.

He then fled from the Anlong Province, near the Thai border, with 'hostages'. They were the nominal guerrilla leader Khieu Samphan and some 'ministers', including Tep Khunnai, the 'territoral integrity minister'.

On 26 July the mystery about Pol Pot's whereabouts was solved. He appeared on film during a two-hour trial for treason carried out by former comrades. Three of his guerrilla comrades were tried alongside him. The show-trial was witnessed by 500 villagers in a settlement deep in Anlong Province. The dramatic footage was shot by American cameraman David McKaige, who was hired by the *Far Eastern Economic Review*'s Cambodia correspondent Nate Thayer, who had been covering the Khmer Rouge for 15 years.

At the age of 69, Pol Pot seemed frail and unwell. Some observers of the Cambodian scene believed that his trial was a ruse by the Khmer Rouge to forge new allies. Khmer Rouge expert Christophe Peschoux said, 'They are trying to make themselves more palatable'.

The Khmer Rouge leadership said that it would not turn Pol Pot over to the international community for crimes against humanity, but that he would be kept imprisoned for the rest of his life. The worry was that he was still respected by the Khmer Rouge rank and file.

Cambodian politics had changed radically by August 1997. Prince Norodom Ranariddh was forced from power and Hun Sen assumed sole power. However, he then sought to appoint Ung Huot as first prime minister. A high-level delegation of 37 officials travelled to China to seek the blessing of Cambodia's king-in-exile. It would take more than the monarch's blessing to bring long-term peace to Cambodia, especially as the US opposed Ung Huot's appointment as 'undemocratic'. Nevertheless, as dual-Cambodian-Australian Ung Huot is attractive to the West as a foreign minister – prime minister.

As far back as *War Annual 1*, in 1984, I reported the death toll at the hands of the Khmer Rouge as 2,000,000 and I subsequently confirmed it in further editions. My information came from Western ambassadors in Phnom Penh, who told me that their governments did not want to accept this figure; it was 'too outrageous to contemplate' and 'too high to be credible'.

References

1. Australian Defence Force sources.
2. Information from Try Chhoun of Adhoo, the human rights group.
3. Cambodian army intelligence is highly professional but too much under political influence for its reports to be accepted without reservation.
4. For my report on the war in Cambodia I am particularly indebted to diplomats. Without exception, they find their dealings with the Phnom Penh government ''confusing and exasperating'. Several point out that 16 years after the Vietnamese army drove the Khmer Rouge into the western hills nobody has been charged with any form of crime. One diplomat, who has now left Phnom Penh, has suggested to me that the Khmer Rouge has paid large amounts of money in bribes to government ministers and officials. It is certainly one of the wealthiest terror organisations in the world.
5. Chhuk Rin was speaking to Michael Sheridan of *The Sunday Times*, whose report was published on 4 August 1996.
6. It is now largely forgotten that the king, when he was prince Sihanouk, in 1970 urged people to join the Khmer Rouge to resist the US-backed Cambodian regime. Many more 'ordinary'' people joined Pol Pot's organisation after Vietnam's invasion.
7. Ranaridd was speaking to Tim Larimer, *Time* Magazine, 23 September 1996.
8. In a 1979 trial, Ieng Sary was sentenced to death for murder and other crimes but everybody from King Sihanouk down seems prepared to forgive and forget.

11

Chechnya: The Great Russian Humiliation

Chechnya, a small Muslim enclave of about 1.2 million people in the North Caucasus wanted, in 1992, to break away from the Commonwealth of Independent States – the rump of the former Soviet Union. President Yeltsin, rather than fight an open war against Chechnya, gave covert military and financial support to Doku Zavyayev and other pro-Moscow opponents of Jokhar Dudayev, the Chechyan leader.

This Yeltsin tactic of creating a puppet government might have seemed a clever ploy because it did not look like Russian power trying to crush a weak and tiny republic. But television crews were on the spot and they showed Soviet-made tanks and Russian soldiers aiding the anti-Dudayev forces. The Russian people, who had been kept in ignorance of their army's involvement, were shocked, especially when they learnt that the Chechnyan rebels had captured 120 Russian troops.

To recover his waning prestige, Yeltsin put pressure on both sides: they must lay down their arms within 48 hours or the Russian army would invade Chechnya with overwhelming force and crush the civil war. The deadline was 15 December 1994. He was alarmed that if the Chechens were successful in their secession attempt more of the 89 ethnic republics and regions of the Russian Federation would also seek independence.

He massed transport aircraft, armour and troops in the neighbouring republic of North Ossetia. In the face of this threat pro-Dudayev and anti-Dudayev fighters came together and volunteers flocked to Chechnya's capital, Grozny, where Chechnya's new chief of staff, Aslan Maskhadov, declared: 'The North Caucasus will become another Afghanistan.'

On 11 December, 75,000 Russian troops invaded Chechnya but to reach Grozny they had to fight their way through Ingushetia, another ethnic Muslim republic that was anti-Russian. From the outset, the war was a debacle for the once mighty Russian war machine and, as it was widely televised in Russia, the public saw the results of the setback. Superb irregular fighters, the Chechens demoralised the Russians to such an extent that some units avoided combat.

Technically, Russian troops held Grozny by the end of April 1995 and Yeltsin ordered that the fighting must be over by 9 May, the 50th anniversary of the end of the Second World War, but the war continued with great loss of civilian life. On 30 July Chechen and Russian negotiators agreed on a disengagement and on demilitarisation of the republic. At this time Russian sources admitted that 1,800 Russians had been killed, 250 were missing – probably butchered by the guerrillas – and 6,000 had been wounded. The fragile peace was broken on 21 September 1995 when the Chechens exploded a bomb under a convoy carrying Oleg Lobov, Yeltsin's special representative in Chechnya. He survived the blast.

With each side guilty of provocation, the authority of the Russian generals and of the Defence Minister, General Krachev, suffered. On 10 January 1996 Yeltsin ordered another assault aimed at bludgeoning the Chechens into surrender. Fire from artillery and helicopter gunships reduced the eastern Chechen town of Novogroznensky to rubble. The stated reason for the bombardment was to save hostages being held by the rebels; in the event many of them died under the Russian shells and rockets. The duplicity shown by Yeltsin was typical of the conduct by both sides in the bloody conflict.

The War in 1996–97

In April 1996 the rebel leader, Jokhar Dudayev, died. According to credible reports, he was assassinated – and in an unusual way. He was using a portable satellite telephone when a rocket attack was made on his position. It would have been possible for Russian military intelligence to have plotted his position. He was succeeded by Zelimkhan Yandarbiyev, who was strongly supported by the separatists' military leader, Aslan Maskhadov. At the end of May, in what was hailed as a 'peace breakthrough', Yeltsin met Yandarbiyev at the Kremlin; Dudayev had vainly sought such a meeting for four years. On 10 June, in the city of Nazran, the two sides signed peace accords which, among much else, stipulated a withdrawal by Russian forces by 31 August. The Kremlin meeting took place just six days before the first round of the Russian presidential elections and it was obvious that Yeltsin's timing of the meeting was an electoral ploy.

Both sides were slow to fulfil their obligations. The Russians promised to dismantle their roadblocks by 7 July but on the contrary most were strengthened. The Chechen command promised to cease firing at Russian checkpoints but continued to do so. On 9 July the Russian army shelled a Chechen village, Gekhi, with heavy weapons. The day long operation, which involved helicopter gunships, was intended to 'liquidate' a group of Chechen fighters but independent observers stated that the only casualties were among civilians. These observers said that they had been given four different versions of how the fighting started. The most likely was that a Russian soldier was accidentally wounded by a comrade but General Vyacheslav Tikhomirov, the hawkish commander of the Russian forces in Chechnya and a critic of the peace process, used this incident as an excuse to blame the Chechens and so begin an onslaught against them. During the Gekhi fighting General Nicolai Skripnik,

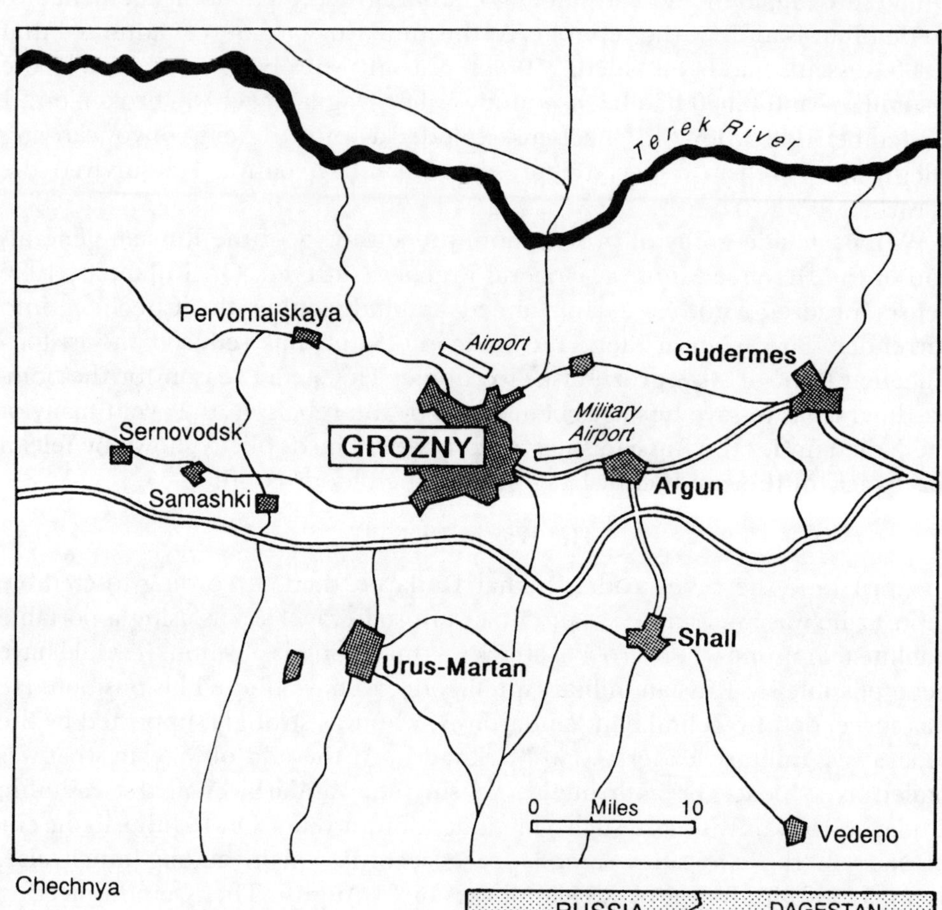
Terek River
Pervomaiskaya
Airport
Gudermes
Military
Airport
Sernovodsk
GROZNY
Samashki
Argun
Shall
Urus-Martan
0 Miles 10
Vedeno
RUSSIA
DAGESTAN
CHECHNYA
INGUSHETIA
GROZNY
DAGESTAN
NORTH
OSSETIA
0 20
Miles
GEORGIA

Chechnya

deputy commander of the North Caucasus region, was killed when his armoured vehicle hit a landmine.

Elsewhere, the village of Makhety was blockaded by Russian troops and pinpoint strikes were made on buildings where Zelimkhan Yandarbiyev's headquarters might be. In response, Yandarbiyev asked his own people for restraint so that fighting would not spread. Most Russian newspapers were now openly interpreting the resumption of fighting as a return to military methods by Moscow, now that the election was out of the way. One commentator, Maris Eismont, wrote: 'The first electoral promise by the president to be broken with incredible ease was the one that attracted most doubters to his side.'[1]

In August, the Chechen rebels took the war to the Russian army. Mounting a counter-offensive, about 2,000 fighters made a dawn raid on Grozny, killing or wounding about 150 Russian troops. They held part of the city's ruins even when the defenders called in helicopter gunships. Simultaneously, the Chechens attacked Russian positions in the towns of Argun and Gudermes. The planner and leader of this set of operations was Shamil Basayev, who already had shown his daring by a surprise attack on Grozny in March 1996.

Aslan Maskhadov, the Chechen chief-of-staff, said that the fresh actions had a single purpose – to show that the war in Chechnya was not yet over. This message was emphasised by a bomb detonated along the main route in Moscow used by Kremlin leaders to reach their offices. In response, Sergei Stepashin, head of the government commission in Chechnya, said that no further talks with the rebels were now possible.

The Russian authorities sealed off all approaches to Grozny and declared a dusk to dawn curfew. The level of trust between the two sides was so low at this time that the only people capable of mediation were the six members of the harassed Grozny Mission of the Organisation for Security and Co-operation in Europe (OSCE), led by Tim Guldimann.

In the fierce street fighting that followed in Grozny the Russian troops were either cut off or driven out. By the night of 8 August the number of Russian casualties had reached 300 and, even worse in a strategic sense, the Russian army had experienced one of its worst humiliations since the war began. At Argun, the guerrillas even beat back a Russian armoured column.

The Chechens were intent on spoiling Yeltsin's swearing-in ceremony on 9 August. One fighter was quoted by *Izveztia* as saying: 'We believed Yeltsin when he said he would stop the war. We not only ceased fighting in our territory but even allowed the elections to be held. But now it turns out that Yeltsin has fooled everyone.'

The rebels were never short of weapons, many of which were bought from Russian officers. Grossly underpaid, some officers made fortunes by selling weapons to the enemy. The guerrillas also captured an adequate number of armoured personnel carriers by shooting at their tyres and so immobilising them.

The latest army commander in the field, Lieutenant General Konstantin

Pulikovsky, lost control of the situation; not that he admitted as much, saying only that the position was 'complex'. Complex beyond all understanding. The few thousand rebels with little more than their personal weapons, beat back 10,000 Russian troops, at least 7,000 of whom were cut off. Some guerrillas laid siege to Grozny's two airports, Khankala and Severny, where many senior officials lived in bunkers.

Yeltsin's language verged on the panic-stricken. Declaring a national day of mourning for the Grozny victims – he did not specify civilians – he said, 'We will crush the terrorist offensive. The attacks will be resolutely suppressed. I will not allow these men to speak to the government in the language of blackmail.'

The situation was farcical. Some Kremlin leaders were ordering the troops out of Grozny while others were ordering them to retake the captured city.

Enter Aleksandr Lebed

By 11 August a humanitarian disaster had overwhelmed Grozny. Thousands of desperate refugees, many waving white flags, fled from the capital, even during artillery barrages, to find sanctuary and to search for food and water. Medical supplies were non-existent. The rebels, flushed with success after wiping out a column of tanks and armoured personnel carriers on the edge of the city, were not withdrawing and the Russian troops could not withdraw.

Now admitting 'gross miscalculations', Yeltsin sacked his old friend Oleg Lobov as the Kremlin's envoy in Chechnya and replaced him with former paratrooper General Aleksandr Lebed, who had long advocated a political settlement with the Chechens. With the backing both of the Chechen leadership and of important men in Moscow he was considered to have a good chance of ending the war. One of his backers was Viktor Chernomyrdin, now confirmed as Prime Minister in the *Duma*, the Russian Parliament. 'Lebed is a military man and used to tackling such problems', said Chernomyrdin. 'I am sure he will cope with the task. He simply must be successful.'

Having fought the presidential election against Yeltsin, Lebed was already a public figure and from his electoral promises the Chechen leadership, through its spokesman, Moviadi Udogov, cautiously welcomed the big and generally blunt former general, now the National Security Advisor.

Lebed is an interesting personality whose enormous ego is likely to be prominent in Russian politics for a generation. He had already participated in most of the significant events involving Russia for a full 30 years. A Cossack, with all that this means in terms of military taste and toughness, Lebed became a cult figure in the army. Some junior officers even tried to imitate his hoarse voice, not knowing that it is the result of an early fight with a rival boy-gang when he was beaten with an iron bar and needed emergency surgery. But he did not spend all his time scrapping and at school was an excellent student. He was present on the momentous day that KGB troops were brought into his town to quell a riot by striking workers; two of his schoolmates were accidentally killed by warning shots fired by the soldiers. Lebed was fascinated by the action of the local military commander in refusing to fire on civilians, as he was supposed to do. Publicly, this officer tore up his Communist party card and shot himself dead.

Lebed served in Afghanistan where he became a legend as a leader. But it is an army folk tale that people talk most about. He had discovered that older soldiers were bullying younger ones, even to the point of torturing them with electric shocks. Ten men were named as offenders but they denied knowledge of this and other abuses. Lebed, who had his own sources of information, ordered the ten to stand at attention in line and then, one by one, he broke their jaws with a single punch from his powerful boxer's fist. Lebed admitted to the action in his book, *Pity the Motherland*, published in 1995. Disillusioned by incompetent politics and military leadership in Afghanistan, Lebed returned home to lead paratroopers against various independence movements.

In 1988 he was under orders to protect the Armenian minority in Baku, the capital of Azerbaijan, where they were being slowly annihilated. He was in an Armenian quarter one night when the electricity was cut off. Lebed recognised this as a sign that a pogram would soon take place. He smashed down the door of the Azerbaijan Communist party's office and demanded an explanation. The power had been cut off for economic reasons, the secretary told him. Lebed took his pistol from his holster, placed it in front of him and ordered that the light be restored in one hour. His first warning, he told the Party leader, would be a pistol whip in the teeth. There would be no second warning. Power was restored within 42 minutes.

In August 1991 hardline Russian communists had tried to regain control of the Soviet Union from the democratic President Mikhail Gorbachev. Having held him captive in the Crimea, the rebels were opposed by the pro-democracy Yeltsin and his supporters, who had taken over the Russian White House. Lebed, with his paratroopers and an armoured column, was ordered to 'secure' the White House, an ambiguous and irrelevant order in the circumstances of the time. The shrewd Lebed did not attack the pro-democracy demonstrators but defensively encircled the building. This done, Lebed strode into the White House to introduce himself to Yeltsin. He advised Yeltsin that in the absence of the imprisoned Gorbachev he could easily win the loyalty of the army by at once declaring himself commander-in-chief of the Russian forces. Yeltsin did this the next day and the coup fell apart.

Promoted to major-general, Lebed commanded the 14th Army in Transdniestr, another autonomous region fighting for its independence from Moldova. Faced with contradictory, hesitant and sometimes stupid orders from Moscow, Lebed, on his own volition, began a pre-emptive artillery barrage against the Moldovan forces. This abruptly ended their war and he became a hero to the public but not to the other generals, many of whom he charged with corruption and inefficiency. He spoke his mind too frankly for his own good, accusing the Kremlin and even Yeltsin of stupidity in dealings with Chechnya. He ensured his dismissal with a comment he made to the media about the defence minister, General Pavel Grachev. 'I don't like prostitutes, whether they are wearing skirts or trousers.'

His resignation from the army – actually he was forced out – did not take Lebed out of the public eye, as his enemies had hoped, but took him into politics. Inexperienced in this field, he made one misjudgment after another and at one time alarmed his western supporters by saying that democracy would only flourish in Russia after a long transition period.

Finally, the desperate Yeltsin gave him the job of negotiating peace with the Chechens. His enemies expected him to fail, perhaps even to be killed, and he was shot at twice as he was driven to his first meeting with rebel leaders in Grozny. He negotiated a brief lull in the fighting and began to repair some of the damage done by several failures to keep promises. Maskhadov described him at the time as the only man in the Russian government not stained with the blood of the Chechen people . . . 'the only man who has a moral right to talk to us and to end the war'.

As usual, Lebed pulled no punches in his criticism of his political chiefs in Moscow. He stated that Prime Minister Chernomyrdin had mishandled the crisis and he demanded that Yeltsin sack General Anatoly Kulikov, Russia's interior minister, as the man principally responsible for the Chechen tragedy. Lebed accused Kulikov of deliberately prolonging the war and of undermining his authority. Certainly Kulikov was leading the large establishment faction which resented the rapid rise of Lebed.

Lebed achieved an uneasy peace. The separatist fighters and the Russians began to disengage their forces and in some sectors of Grozny rebels and Russians engaged in tentative conversation. Even General Tikhomirov kept an appointment for talks with Aslan Maskhadov.

Lebed left Grozny feeling triumphant but his reception in Moscow was cold. He demanded a face-to-face meeting with the President in order to tell him what he had done in his name but it was refused. Lebed needed to return to Grozny but he could not take with him Yeltsin's endorsement. Indeed, in radio interviews Yeltsin was critical of Lebed's peace mission. The major newspaper *Moskovsky Komsomolets* was probably speaking for many people when it stated in an editorial that the Russian military had been beaten and humiliated by the rebels and that General Lebed's peace agreement was nothing more than 'the terms of the capitulation'.

The truce, signed by Lebed and the rebels' commander Maskhadov, called for a Russian withdrawal from Grozny and an end to hostilities across Chechnya by noon on 23 August. Since the start of the month 406 Russian soldiers had been killed, 1,264 wounded and 130 were missing presumed dead. 'Federal troops will withdraw from Chechnya because constitutional order cannot be established by air raids and artillery shellings', Lebed said as he signed the deal.

Lebed's peace held, even though he himself was dismissed from his post as special negotiator. His time would come. In the meantime, the leader of the

Chechen coalition government is Zelimkhan Yandarbiyev and the Prime Minister is Aslan Maskhadov. Moscow's puppet leader, Doku Zavyayev, fled to Moscow.

Shamil Basayev – Folk Hero

The most renowned Chechen guerrilla leader, Shamil Basayev, has been mentioned several times in this account of the war but he was always a shadowy figure, rarely photographed by the media and giving only the occasional brief interview. He became prominent as the war ended, by which time he was a folk hero for his exploits against the hated Russians.

Some of them were criminal. He rounded up more than 1,000 hostages in a hospital at Budyonnovsk, southern Russia in 1995; more than 100 of them died. He hijacked an aircraft flying from Russia to Turkey, and he robbed banks.

To the anger of every minister and official in Moscow, Basayev announced that he would run for the presidency of Chechnya, but Basayev hates Russians and their anger would only please him. At one of his first media interviews in Grozny after the peace accord had been reached he said: 'Perhaps we should restate Churchill's three rules about Russians – don't believe the Russians, never make friends with the Russians and never let a Russian into your cow shed'.

The bald, heavily-bearded, 31-year-old who was several times wounded has a remarkable influence over younger Chechens who see him as a hero. His features are seen in homes and public places all over Chechnya, and their presence shows that neither Basayev nor Chechnya feels defeated. He told reporters: 'It will be hard for the Russians to work with me. I won't let them rob Chechnya. I will make sure that all agreements with Russia are first and foremost in the interests of Chechnya'.[2]

Chechnya in 1997 was an extraordinary place if only for the irony that it was more dependent on Russian money than ever before. Chechnya itself, lacking an administrative structure, had no money to staff the hospitals and schools, pay pensions, repair the oil refineries and rebuild two large cities and possibly 1,000 wrecked villages. The ordinary people were at the mercy of the warlords and the Chechen mafia, who recovered their Mercedes from neighbouring Ingushetia as the conflict ended and drove about Chechnya to display their power. Abduction at once became a growth industry, with the equivalent of £25,000 being demanded in ransom.

The only structure of government working in December 1996 was the Joint Central Commendatura, under the charge of Colonel Valery Dragoshenski of the Russian Ministry of the Interior.

Russian Stubborn Refusal to Learn

The Russians, politicians and military leaders alike, have learnt nothing from the military history of their own country. Their involvement in the war against the Chechens and especially the Chechens' offensive of 1996 proves this conclusively.

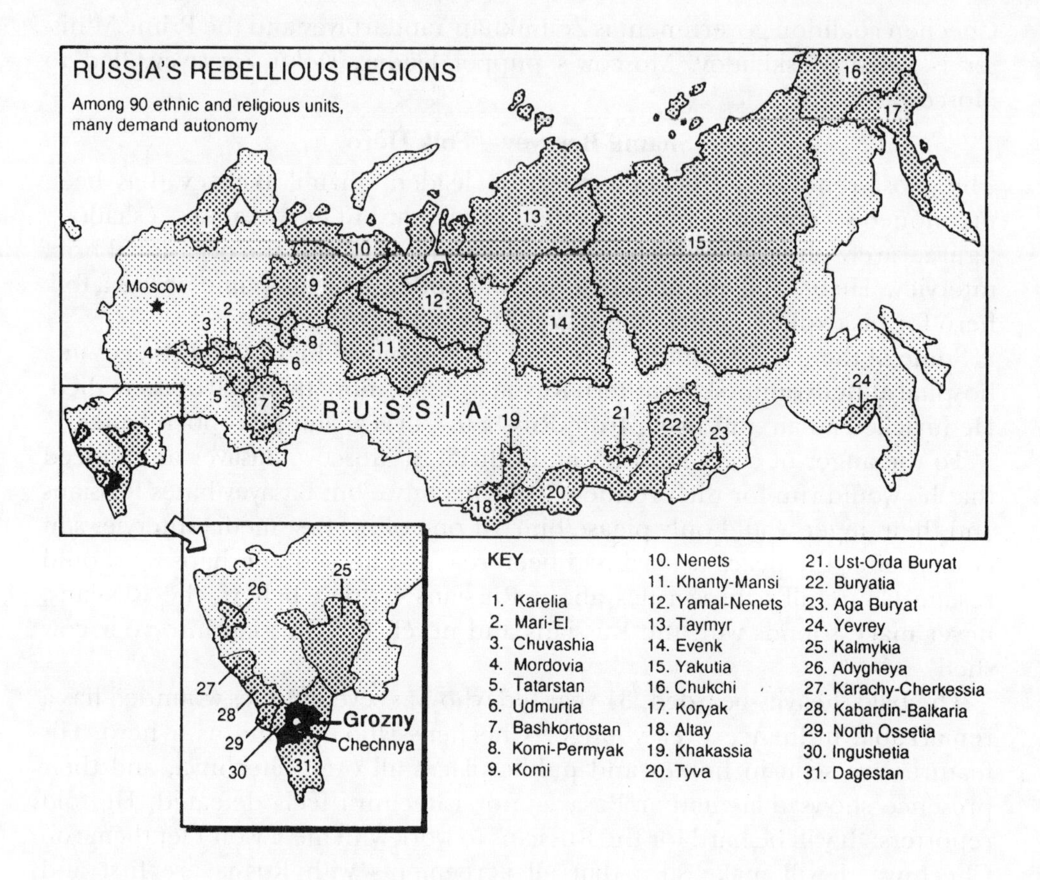

To find a point of reference for proving the Russian neglect of historical lessons it is instructive to go back as far as the 1820s, when General Alexei Yermolov was sent to subdue Chechen guerrillas in the Caucasus mountains. He succeeded in turning the entire Chechen population against him. His army demonstrated only that the Chechens were unconquerable.

Twenty years later a Russian prince, Aleksandre Chernyshev, had learnt something from General Yerolov's blunders and, as Minister of War for Tsar Nicholas I he produced a thoughtful report in which he said:

> Operations against the mountaineers [referring to the guerrillas of the Caucasus] do not achieve the government's aim of pacifying them. On the contrary, the almost uninterrupted Russian failures boost the mountaineer's morale, while our soldiers become disorganised, dispirited and exhausted.

The Chechens (like other tribes and clans of the Caucasus) were superb horsemen and skilful scouts and ran rings around the much greater forces of the Russians who were hampered by artillery and baggage trains. They were often ambushed and the victims of surprise attacks – a similar fate to that of the

Russian units deployed in Chechnya since the beginning of the war in 1994.

The modern Russian military commanders appear to have no greater knowledge of the Caucasus than their predecessors; they do not know its internal politics and its society and they underestimate the military knowledge of the guerrilla leaders, such as Shamil Basayev. In the earlier campaigns the army had relied on young conscripts, in the mistaken belief that they could easily defeat the ignorant peasants of Chechnya. When this was proved wrong the High Command replaced most of the conscripts with regular soldiers. However, like the conscripts before them they were given no briefing about their Chechen enemies. Poorly paid, these troops had no idea why they are fighting and many secretly sympathised with the Chechens.

Because the soldiers are so ineffective, the High Command uses air power and artillery. The object is to cause great damage, which the politicians in Moscow equate with success, and to minimise their casualties in infantry and tank units. Again these are the failed tactics of previous generations. Of course, no air power was available to the Tsars but they had great siege trains with which they pounded Chechen settlements.

The term 'body count' might be a new one but the practice was common in Soviet imperial campaigns. The general who could report 20,000 enemy dead was applauded and decorated; nobody asked him how many of the dead were totally uninvolved civilians, including women and children. In its report from Chechnya 1995–97 the army did not distinguish between civilian and military casualties.

Another ploy unchanged from earlier centuries was that of creating a 'police force' to control an occupied city. The thinking behind this was that guerrillas were less likely to attack policemen than troops and that fewer soldiers were required in vulnerable garrisons. To the remarkably naive Russian commanders, this policy seemed to be working; they did not realise that most of the police were aiding the guerrillas in many ways, even to the point of telling their leaders the best way to overcome a Russian position. In at least one case a 'policeman' attracted the attention of a Russian tank crew while a party of guerrillas set fire to their tank.

The basic lesson that remains unlearned among Russian leaders is that which Prince Aleksandre Chernyshev did learn in the 1840s. Neither the Chechens nor any other similar group of people in the Caucasus can be defeated militarily. In contrast, the Chechens know that with an occupying army of 40,000 in their country they cannot permanently hold Grozny or any other city. This is why, with the arrival of strong Russian reinforcements they vanished into the hills. Unlike the army's leaders, Basayev and others know their limitations; more significantly, they also know the Russians' limitations.

A report in *Jane's Intelligence Review* of mid-August 1996, which was co-authored by Charles Blandy of the Conflict Studies Research Centre, Sandhurst, doubted that the Russian army, even after the lessons of Chechnya, would improve its performance in the future. The report referred to technological failings; for instance, the much-vaunted T-72 tank had proved vulnerable to determined infantry attack. Penetrating missiles could ignite the tank internally. Bad weather had prevented many Russian aircraft from operating, notably the SU-25s. The report emphasised what many other observers and commentators had referred to – that in time-honoured fashion the Russian leaders had underestimated their opponents.

A fundamental fault had been the failure of logistical support for the combat units. The *Jane's* report mentioned some units as competent and effective, including the VDV (the airborne forces), the 74th Independent Motor Rifle Brigade, the 506th Regiment of the 27th Motor Rifle Division and naval infantry.

Russian Incompetence – An Example

Before being sent to Chechnya early in 1995 the 245th Motor Rifle Regiment had a strength of only 172. In just ten days its establishment was boosted to 1,700 officers and men, most of them raw recruits, and ten days later the unit was posted to Chechnya. This was contrary to army practice, which stipulated that recruits should receive a minimum of three month' and preferably six months training. On top of this, the ill-advised system of rotating officers in Chechnya every three months meant that the 245 Regiment's officers had insufficient experience with their new troops.

On 16 April 1996 the 245th Regiment, which was escorting a supply convoy, was moving through a mountain defile near Yarysh-Mardy when guerrillas ambushed it. Surprise was complete and the regiment suffered grievously. Seventy-three men were killed, 52 wounded and an enormous amount of equipment was lost, including a tank and six APCs. The Russians, shocked and disoriented, fired back wildly but it is doubtful if a single Chechen guerrilla was hit. Yarysh-Mardy was one of the classic ambushes of history.

Such a huge disaster could not be hushed up and General Lev Rokhlin, chief of the *Duma*'s defence committee, was ordered to investigate. He was a good choice, having commanded troops in Chechnya, where he captured Grozny in January 1995. He had been elected to parliament in December 1995. His report was published, apparently in full, in the daily *Nezavisimaya Gazette* on 27 April 1996. The field commanders were guilty of tactical illiteracy, poor communications and low alertness, Rokhlin reported. No officer on the spot had acquitted himself well and the men, being ignorant and lacking unit morale, became a rabble. But his strongest criticism was directed against the chiefs of the defence ministry, which had been 'unconscientious'. Furthermore, reductions in the Russian armed forces – from 3.5 million to 1.7 million between 1991 and 1996 – had destroyed most of the specialised units, such as combat engineers and rapid-assault battalions. The 245th Motor Rifle Regiment, Rokhlin stated, had few skilled riflemen and grenadiers and no sniper specialists.

Trenchantly, the general went on: 'The army has been left without finance. It cannot survive when soldiers do not have enough food, when they are unpaid for months at a time and when new equipment is not reaching them.' According to the report, the army had not received a single new tank in a full year. Rokhlin's recommendations were:

- A sound, modern military doctrine was essential.
- Not less than 5 per cent of the gross national product should be spent on the armed forces.
- A unified command body should be set up. [Presumably this was to do away with the many military fiefdoms that exist in the Russian forces.]
- A public relations campaign was necessary to impress upon the nation the prestige of military service.

Clearly, General Rokhlin was going a long way beyond his brief, which had been to investigate a single ambush, but his report was accepted without much demur. But this was no guarantee that his recommendations would be accepted.

However, early in December the commander-in-chief of Russia's army, General Vladimir Semyonov, aged 56, was dismissed. Semyonov himself broke the news to Russian journalists, showing them a letter from the defence minister, Igor Rodionov, stating that he was losing his post because of 'actions incompatible with his duty'. President Yeltsin himself had countersigned the letter. It was generally considered that Semyonov, who was not implicated in any of the numerous army corruption scandals, was paying the price for the army's humiliation in Chechnya.

References

1. In *Segodnya*, 10 July 1996.
2. *The Independent*, London, 5 December 1996; an article by Phil Reeves, who was present at the media interview.

I am particularly grateful to a Chechen journalist for much of the information from within Chechnya. He telephoned, faxed and wrote to me from bases in Chechnya and from the neighbouring state of Ingushetia. Sometimes he ran a risk in maintaining contact with me, especially during the period when the Russians were trying to prevent damaging information from getting out of the stricken country.

12

Colombia: A Madhouse of War

Background Summary

Colombia has been infested with paramilitary organisations since the late 1950s when a struggle began between groups labelling themselves, respectively, 'liberals' and 'conservatives'. In a rational and moral sense these adjectives were quite unwarranted. The most important groups have been identified and described in earlier editions of *War Annual* but they need to be mentioned again in order to make sense of more recent activities. They are:

- Colombian Armed Forces or *Fuerzas Armadas Revolucionarias* (FARC) the military wing of the Communist Party, whose leader until 1996 was Pedro Antonio Marin or 'Sure Shot', at the age of 68 one of the oldest guerrilla leaders in the world.
- 19th April Movement, always known as M-19, the most powerful group until displaced by the Army of National Liberation.
- Army of National Liberation (ELN), a pro-Cuban group. ELN lost its separate identity in 1986 when it merged with the National Guerrilla Co-ordination (CNG), a bizarre kind of guerrilla co-operative. In 1996 it resumed activities under its original name.
- The Patriotic Union, a right-wing organisation of government officials and supporters. It was supposedly a self-defence group but it deteriorated into 'Death Squad B'. The 'B' label distinguishes it from Death Squad A, which was for long considered a 'legitimate' formation.

The narcotic barons created private armies that made the wars of the 1980s even more vicious and complex. Pablo Escobar, the most notorious baron, declared 'total war' against the presidency of Virgilio Barco. He was killed in a shoot-out in Bogota in 1993 at the age of 44.

The government had other triumphs and M-19 and FARC lost so many members that in 1990 they disbanded, though in FARC's case only temporarily. In April 1990 Carlos Pizaro of M-19 was assassinated and in July FARC's 'Sixth Front's' commanders, Miguel Pascuas, was shot.

ELN was active throughout 1995, frequently dynamiting oil pipelines and engaging in widespread kidnapping, partly to intimidate the government and its supporters but also to raise money through ransoms.

In July 1995 the authorities captured Jose Santacruz Londono, whose cartel handled 80 per cent of the world's cocaine traffic. This led to other arrests.

However, President Ernesto Samper himself was accused of links with the drugs cartels and one of his ministers resigned over allegations that he had taken $6 million in 'election contributions' when he was Samper's campaign manager. With the government now vulnerable, the guerrilla groups rejected Samper's peace proposals and began a new offensive.

The War in 1996–97

That a serious situation had developed in Colombia was made clear in March 1996 when the US, accusing the Colombian administration of sabotaging the campaign against the drugs traffickers, suspended aid funds. This followed President Clinton's official notification to Congress that Colombia (among other nations) was not co-operating in the international counter-narcotics fight. Clinton's action had serious consequences for it meant that foreign assistance was radically cut and that the US government had vetoed lending by six major multilateral development banks.

Apart from Colombia, Afghanistan, Burma, Iran, Nigeria and Syria were all denied aid. Another 25 countries, though they produce or tranship drugs, continued to receive aid because they were considered to be making an effort to combat their drugs syndicates. The Department of State assistant secretary for international narcotics, Robert Gelbard, said:

> The decision to deny Colombia certification [as an aid nation] was not made lightly but it is crystal clear that narcotics interests have gained an unprecedented foothold in Colombia. There is no doubt at this point that the administration of President Ernesto Samper receives significant financial aid from Colombian drug lords.[1]

The cut in funding hit Colombia hard because at a stroke it meant that no further US counter-insurgency aircraft would be supplied.

Apparently in response to the American embargo, President Samper announced 'emergency powers' to fight both drug-related crimes and the rebels. In mid-April the rebels struck back, as is their custom when they consider themselves 'provoked'. An estimated 150 FARC rebels attacked a six-vehicle army convoy with explosives, machine-guns and grenades about 400 miles south of Bogota, near the border with Ecuador. In this region the army routinely guards oil installations which are frequently attacked by FARC. FARC and other rebel groups intensified their attacks.

On the same day as the ambush, a Colombian House of Representatives peace envoy, Liberal Party Representative Josez Maya, was kidnapped by the ELN, whose rebels attacked a police station at Medellin, wounding four policemen. In the southern ambush 31 soldiers were killed but only three rebels.

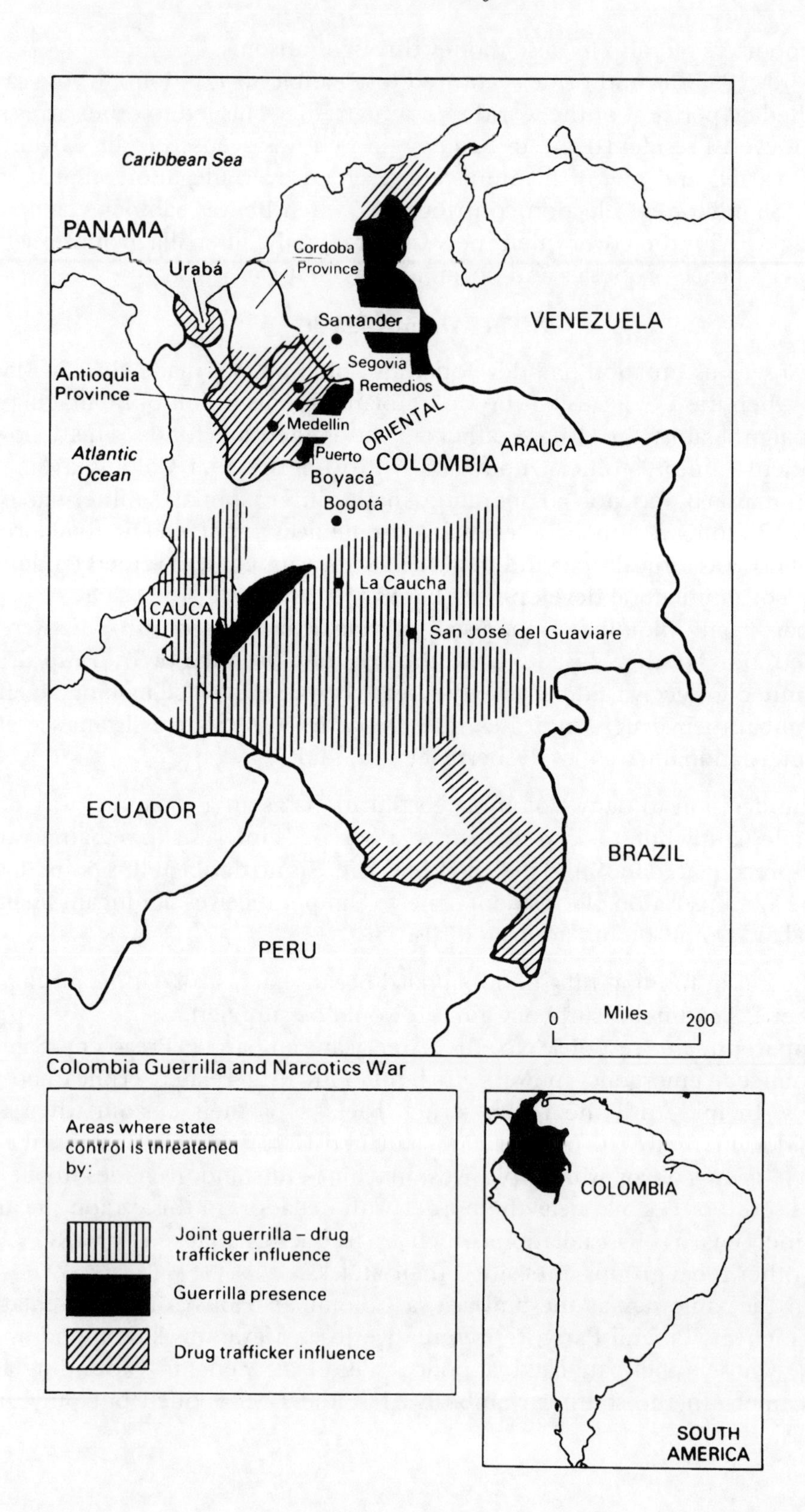

Colombia Guerrilla and Narcotics War

FARC kept up its 'offensive'. In mid-September its fighters attacked the small agricultural town of Mocoa, on the edge of Colombia's southern Amazon region. They dragged the bullet-ridden body of a young soldier into the square, set it on a seat and pinned a message to it which read: 'The Revolutionary Armed Forces of Colombia are back. We will fight to the end to bring down the corrupt government of Ernesto Samper.'[2]

They then blew up the town's power plant and the modern offices of the regional government. With the unpaved airstrip on the outskirts of town constantly under fire, the only access to Mocoa was by bus, which had to stop at a checkpoint manned by guerrillas. Armoured patrols ran the gauntlet down the road but as usual the rebels filtered into the jungle when this happened.

The Mocoa attack was only the first of many in September. FARC and ELN had regrouped their estimated 18,000 men, a move that took the government by surprise. The rebels actually captured smaller army bases in the southern region of Putumayo, Guavaire and Caqueta, killing 150 soldiers and 50 civilians – so called collaborators. The attack on the Putumayo base involved about 400 rebels armed with machine-guns, grenades and mortars. At Las Delicias they took 60 soldiers hostage and little hope was held out for their lives.

Alfredo Rangel, a government spokesman, said: 'The guerrillas have taken control of vast areas and forced our security services to take defensive action. Our units were unprepared for the assault.' This was an astonishingly frank and alarming confession of military ineptitude.

Towards the end of September FARC and ELN groups were in control of most of the remote Amazonian lowlands and were making forays into the highland areas close to the country's main cities.

According to army reports, the guerrillas were retaliating against recent government efforts to destroy coca-leaf crops, used in the making of cocaine. These efforts were a direct result of the American aid embargo. The coca farmers violently protested and the guerrillas – as many times before – exploited the protests. Juan Carlos Esquerra, the Defence Minister, described the resumption of warfare as 'particularly worrying'. He said: 'By moving into the cocaine-producing areas the guerrillas can take control of the drug trade, acquire much more money and spend it on sophisticated weapons'.

Many villagers who are caught in the middle of the war nevertheless support the guerrillas. They consider that FARC is defending their own cause – that of breaking out of their endemic poverty. The oil fields are exhausted, the government wants to burn coca crops and there is no new industry.

Leader of the late-1996 FARC offensive was Jorge Briceno, successor to Pedro Marin. Aged 40, he does not look like a military commander though he wears combat fatigues and a military style blue beret. The product of thorough training by FARC rebels now no longer involved in field activities, Briceno is known to be lukewarm in his Marxist beliefs. He has told Colombian reporters that he 'simply' wants to bring down the Samper government so that a less corrupt one can take its place.

The government is largely disgraced because it is now beyond doubt that the Cali cocaine cartel financed its 1994 election campaign, but it is not weak. Even the resignation of vice-president Humberto de la Calle in protest at the corruption did not affect the government's position.

New Military Chiefs

The fear that FARC and ELN would take over as Colombia's leading drug syndicate came to a head in November 1996 when President Samper appointed a new commander-in-chief of the armed forces and new chiefs of the army, navy and air force. The new overall chief is General Harold Bedoya; General Manuel Jose Bonett took over the army, Admiral Roberto Serrano the navy and General Fabio Zapata the air force.[3] Rebel leaders claimed that the sweeping new appointments indicated that the government had panicked.

Pressure came not only from the rebels and from the Clinton administration, but also from Amnesty International which in November 1996, claimed that US military equipment, intended for anti-drug operations, had been diverted to counter-insurgency units which had been responsible for the deaths of many thousands of civilians. Indeed, the military has probably killed more civilians than the rebels have; entire villages have been wiped out because a few people there were believed to be rebels.

Amnesty demanded an immediate halt to the $40 million US military assistance – which had not been affected by the presidential aid embargo. The organisation's spokeswoman said that the US government must produce a plan to end 'abuses' and must account for the past 'misuse' of US assistance.

Amnesty's intervention alarmed the Samper government. The Colombian embassy in Washington stated: 'We deplore the call for suspension of American military assistance. This assistance is used primarily in support of anti-narcotics operations for transporting troops, protecting aircraft and providing humanitarian aid.'

According to the Colombian Attorney General, in November 1996 his department was investigating 1,300 allegations of violation of human rights made against members of the security forces. Observers say that it is reasonable to assume that no action will be taken against anybody – 'for lack of evidence'.[4]

Meanwhile, in December the interior ministry began a programme to encourage respect for human rights. Ministry publicity said that it was hoped to influence the rebels as well as the national services. Clearly, among the rebels the Ministry was aiming its publicity at the young women who belong to FARC and ELN. Many of them lead patrols.

Guerrilla Warfare is a Profitable Business

During 1997 the most publicised sufferer at the hands of the ELN and FARC was Occidental ('Oxy') the American company, which replaced BP as the oil company which had its assets most frequently hit. On 6 July, guerrillas hit the line from Oxy's Cano Limon field to Covenas, the west coast Caribbean oil port. It was the 36th attack in 1997. When the army flew in troops to protect the repair engineers the guerrillas shot down their helicopter killing 24 soldiers and officials. On 10 July, the guerrillas killed ten more soldiers on pipeline protection duty and Oxy suspended production.

Executives sometimes say that they might as well abandon the Colombian oil enterprises for good. By law, all the oil companies pay an official protection or 'war tax' of $US1.25 a barrel. BP pays even more than this. In November 1985, it agreed to pay the army a further $US5.4 million. It hardly gets value for money.

According to the US-based Human Rights Watch, Royal Dutch Shell pays the army $2.7 a year in secret deals to protect the Arauca oil fields. In a report, Human Rights Watch says:

> That oil facilities and oil company personnel need to be defended from guerrilla attack is not disputed. The question is how a company like Shell, that claims to express support for fundamental human rights, conducts itself.

Shell, with other companies, is accused of aiding and abetting the army in its oppression of ordinary people during its efforts to defeat the guerrillas.

The guerrillas are unworried. The army has 11,000 of its 80,000 men guarding oil installations, but ELN and FARC strike where they please. The small foreign companies (logging, drilling for oil or mining) are even more at risk than the oil giants. When the local guerrillas demand protection money – they generally refer to it as 'profit sharing' – the foreign company might approach the mayor for advice and help. Even if a mayor is not in league with the terrorists he is scared of them. In the first six months of 1997 54 mayors had been murdered.

The irony is that the major guerrilla/terrorist groups are left-wing champions of the common people whom they want to liberate from the capitalistic clutches of the political right. However, the guerrilla leaders have healthy bank accounts abroad and much of the money fed into these accounts is from extortion. Many smaller foreign companies do much to help local communities, perhaps by building schools, health centres and, of course, by employing local people. All this, it seems, is exploitation – or so the guerrilla chiefs say.

It is difficult, in this violent country, to tell truth from propaganda. The oil companies operating in Colombia are often attacked as ruthless abusers of human rights. The truth may be that they are the victims and the subjects of expensive smear campaigns. They will certainly not be closing down. Enormously fruitful fields are coming on stream, such as BP's Casanare field, producing 500,000 barrels a day at the end of 1997. The guerrillas will want a share of the spoils – and so the cycle continues.

References

1. A statement by Gelbard to a US Congress committee on 5 March 1996.
2. Reported by Gabriella Gamini, a Colombian journalist noted for her refusal to be intimidated by either the government or the rebels.
3. Reported by *Jane's Defence Weekly*, 13 November 1996.
4. Because of intimidation, witnesses can never be found for a prosecution to proceed.

13

East Timor Resistance War

A better label for this war might be Indonesia's War of Oppression, though some supporters of the oppressed East Timorese people would say that Indonesia's policy seems to be that of creeping genocide. In 1991 the Roman Catholic Archbishop of East Timor, Carlos Belo, said that the East Timorese were 'living a time of terror'.

The problems for East Timor began in 1975 when the Portuguese abandoned it after 500 years of colonialist rule. This precipitated a civil war, which was won by the *Frente Revolucionara de Timor-Leste Independente,* commonly known as Fretilin. Fretilin's leaders declared East Timor to be an independent state but Indonesia, which already possessed West Timor, annexed East Timor and named it Irian Jaya, the 27th state of Indonesia. Within 10 years it was widely believed that the Indonesian army had massacred 200,000 of the 1975 population of 680,000.[1]

It might seem strange that an occupying army should embark on such a level of killing against inoffensive unarmed people but the troops were inspired by a perverted religious element. Indonesia has the world's largest Muslim population – verging on 200 million. The East Timorese people are Roman Catholics and the Muslim soldiers of Indonesia were told that they were embarking on a *jihad* or holy war against Christian infidels.

Operating from bases in forests and mountains, a Fretilin resistance 'army' – Fantilin – came into being. At first primitively armed, the guerrillas raided army camps and posts to equip themselves for what would obviously be a long war. The army carried out bloody reprisals and there was much evidence, especially from Amnesty International, that many East Timorese were being tortured.

Another tactic employed by the Indonesian authorities has been to force hundreds of thousands of people out of the mountains into the cities and villages on the malaria-infested plains which are much more easily policed.

The local Tatum language has been banned and Bahasa Indonesian imposed by force. Roman Catholicism is aggressively discouraged.

While the UN has always condemned Indonesia's invasion of East Timor, powerful foreign governments, notably Britain, the US and Australia, for their own economic reasons, have never supported sanctions against Indonesia.

Indeed Australia is the only country in the world to recognise the Indonesian annexation of East Timor and recognition was soon followed by a treaty on the exploitation of oil beneath the East Timor Sea. This angered Australian veterans' organisations, which recall the heroism of the Timorese during the Second World War who often put the safety of Australian troops above their own during the Japanese occupation.

Most governments, including Australia, generally give sanctuary to East Timorese dissidents. For most ordinary Australians the military rule in East Timor is outrageous and the murder there of six Australian journalists in 1976 has not been forgotten.

The Pope showed his support for the unfortunate East Timorese when he made a brief visit in 1990. Almost immediately after his departure the army crushed popular demonstrations, most of them nothing more than effusion of joy over the Holy Father's interest in the people of such an insignificant part of the world.

The Dili Massacre

On 28 October 1991 a young Timorese independence activist, Sebastian Rangel, was shot dead by troops in Dili church compound. Several hundred mourners gathered at this church at 6.15am on 12 November for a memorial mass and then marched peacefully through the town to where Rangel had been buried. When troops opened fire on the gathering more than 90 people were killed at once and others later succumbed to their wounds.

A cameraman from Yorkshire Television filmed much of the ghastly event and his film was smuggled out of the country to give the lie to Indonesian denials of the carnage. Then the commander of Indonesia's armed forces, General Try Sutrisno, publicly defended his soldiers. 'Come what may', he said, 'let nobody think that they can ignore Indonesia's armed forces. In the end they will have to be shot down.'

The Dili massacre provoked international outrage as well as welcome support for Fretilin but the organisation suffered a severe setback when its leader, Xanana Gusmao, and his deputy were captured.

The Indonesian government's relatively new policy is to swamp the local populace with immigrants from other parts of Indonesia, such as Java, Sumatra and Sulawesi. More than 100,000 ethnic Javanese have been settled on stolen land.

A view commonly expressed in Australia is that the Indonesian government, from President Suharto down, realises that East Timor is an embarrassment and that there is enough space for the burgeoning population without it, but that to grant it independence now would provoke similar claims for independence by other ethnic groups within Indonesia's empire.

Xanana Gusmao's Successor

With Gusmao languishing in prison, the Indonesian authorities believed that the Fretilin resistance would wither and die but within a year it had a new leader, Nino Konia Santana, from Tetusla, on the eastern tip of East Timor.

Santana was just 21 and had qualified as a primary school teacher when the Indonesians invaded his homeland in 1975. A practical, commonsense leader, Santana drew much of his inspiration from Gusmao and like him realises that the Indonesians cannot be defeated militarily. The task of the guerrillas is rather to prevent them from integrating East Timor into Indonesia. Put another way, the resistance must prevent the Indonesians from winning and keep going by 'political mobilisation'.

Santana was faced with immense difficulties after the capture of Gusmao and his deputy. The army made many sweeps and arrests and Santana had to restore his followers' confidence in themselves. Unintentionally, he was helped in his task by the brutal tactics of General Teo Syafei, who ruthlessly pursued the guerrillas and their sympathisers. A more intelligent, compromising approach might have produced defections in Fretilin's ranks but Syafei achieved just the opposite – more Timorese developed a hatred for the Indonesians and joined the Fantilin guerrillas.

Some of these are veterans who had earlier surrendered to the Indonesians, only to be roughly treated. Led by a school-master, Antonio Suarez, they returned to join Santana and in several minor encounters they outwitted and outmanoeuvred the army's Red Beret commandos. The Suarez group and a further 70 younger recruits, under 'Commander' Ernesto, opened up areas that Fretilin had abandoned in 1991.

Santana gave details of Fantilin's strength to a visiting British film-maker, Max Stahl.[2] At the place where Stahl met Santana there were about 800 guerrillas, armed with 500 weapons captured from the Indonesians. They were divided into independent companies and guerrilla units. Backing these full-time fighters in Santana's own region were another 400 men while a further 500 could be summoned by radio or runner. Behind these frontline fighters were tens of thousands of activist and sympathisers.

Some weapons do reach Fantilin from outside East Timor but most have to be captured in combat. Nevertheless, there is a steady trickle of supplies from corrupt – or even sympathetic – officers in the army. Santana and his lieutenants claim effective control over large areas of the mountains that form the spine of East Timor. Santana told Max Stahl:

> Indonesia only takes control during a military operation. After the operation is over they lose that control and we are able to carry on with our lives. We meet to solve problems, evaluate our actions and reorganise.

Stahl and other Westerners who have penetrated into Fantilin territory – and with the right contact in Dili and elsewhere this is not difficult – find the guerrillas and sometimes their families living without fear. Their intelligence system is better than the army's and they generally have plenty of warning before an enemy military operation. Scout helicopters are of little help to the army in this mountainous and jungle-covered terrain.

It is interesting that compared with similar insurgents elsewhere in the world, the East Timorese have never engaged in terrorism outside their own borders; there is no equivalent among East Timorese to the IRA, the PLO or ETA. Instead of resorting to violence the younger East Timorese join in loose organisations that set out to inflict maximum diplomatic embarrassment on the Indonesian government and the foreign governments which collude with it. In 1996–97 scores of Timorese seeking political asylum scaled the walls of Jakarta's embassies. These 'escape' attempts invariably result in international publicity.

The Nobel Peace Prize

In October 1996 the plight of the East Timorese was dramatically and internationally highlighted when the Nobel Peace Prize was awarded jointly to Bishop Carlos Belo, aged 48, and to journalist Jose Ramos-Horta, 51, for their efforts to free the country from its invaders. President Sampaio of Portugal, which champions the East Timorese cause, said: 'The award reflects their indefatigable work in the service of human rights and peace in the territory'.

Bishop Belo had demonstrated great courage from his cathedral in Dili, trying to mediate between a brutal military regime and the resentful and militant Fretilin. Ramos-Horta, spokesman and external leader of Fretilin, travels extensively trying to mobilise support for independence. From his base in Sydney he said the prize should have gone to the imprisoned Xanana Gusmao, 'an outstanding man of peace and democracy, a man of courage'. Predictably, the Indonesian government issued yet another angry denunciation of Ramos-Horta.

The timing of the award could hardly have been more embarrassing for the Indonesian government. Four days later, on 15 October, President Suharto arrived in Dili with all the show of an emperor on a previously arranged 'royal' tour. He had come to view a huge statue of Christ recently built on a rocky outcrop on the coast. It is supposed to be a symbol of reconciliation from the government to the mainly Roman Catholic East Timorese. Bishop Belo had already condemned the statue as irrelevant to the real needs of the populace.

Ordinary people were able to be more blunt in their criticism. They said that the statue was an intended bribe, while diplomats in Jakarta say cynically that it is supposed to make the people more friendly towards their torturers. The statue is also unpopular with local government employees who paid for it with compulsory deductions from their salaries.

Bishop Belo was among the assembled dignitaries for the 'presentation' of the statue but he was given no part in it. The Indonesian Minister for Religion, for the sake of the visiting media, offered him a few personal words of congratulation on the Nobel Peace Prize and the commander of the armed forces wordlessly shook his hand in passing. According to Belo, the military has twice tried to assassinate him, in 1989 and 1991.

There may be some significance in that President Suharto was making his first visit to East Timor in eight years. A diplomatic source tells me that the president 'detests' the place.

The award of the Nobel Prize to Belo and Ramos-Horta resulted in hundreds of editorials around the world, bringing the kind of publicity to the plight of the East Timorese that cannot be bought. A few extracts from an editorial in *The Times*, London, are typical of the general reaction:

> Nobody can pretend that the award of the Nobel Peace Prize to two courageous men from East Timor will end that land's occupation by Indonesia. Nobody can pretend, either, that the Indonesian government will be moved or shamed – or even very slightly embarrassed – by this latest international spotlight on its continuing brutality . . .
>
> The facts are naked and eloquent. Indonesia marched its army into East Timor, then a Portuguese colony, in December 1975. It was an act no more lawful than Saddam Hussein's forced annexation of Kuwait [1990] and General Galtieri's aggressive misadventure in the Falkland Islands [1982] . . .
>
> According to Bishop Belo, 'All the Timorese want to do is to sit at the table with the Indonesian government and negotiate a peaceful exercise in self-determination'.
>
> Mr Ramos-Horta said, with impeccable modesty, that the Peace Prize should have gone to Xanana Gusmao who is serving a 20-year jail sentence for 'conspiracy to set up a separate state'. Perhaps he is right but the two men honoured are Mr Gusmao's co-conspirators. In fact, it is virtually impossible to find a Timorese man or woman who is not guilty of that conspiracy too, whether in speech or thought or action. Let the world take notice of that, as the Nobel Committee has done, and applaud the bravery of this embattled people.[3]

In an interview with the German news magazine *Der Spiegel* Bishop Belo was quoted as saying that the Indonesian troops in East Timor were treating the people like 'mangy dogs and slaves'. This was not the first time that the bishop had used such strong language in describing the behaviour of the Indonesian armed forces but this time, in the wake of the award of the Nobel prize, his comments angered senior military officers and cabinet ministers. The foreign Minister, Ali Alatas, said: 'We object to his engaging in politics. That is not what the Vatican expects him to do in his position as bishop of Dili.' But Bishop Belo declared:

> What I say publicly in my interviews does not necessarily refer to my personal views nor my personal experience as one Timorese nor as one priest, but rather as the one in whom people confide deeply. Since 1981,

the year of my return from Portugal, until now they pour out their sorrows and pains.

The Timor Human Rights Centre, which is based in Melbourne, issued a report in December 1996 alleging that abuses against the people were still high. Ten East Timorese had been killed in extra-judicial executions between January and September. The report stated:

> The year 1996 has seen a high level of arbitrary arrests, particularly of young East Timorese, and systematic torture of those arrested. There are unfair trials, acts of intimidation and terror, and disappearances. All this highlights the need for the Indonesian government and governments worldwide to address the human rights situation.

According to the Centre, on 10 June 1996 a peaceful protest turned into a riot after Indonesian soldiers shot at demonstrators. The troops fired live ammunition instead of the plastic bullets which the army had been instructed to use following the Dili massacre of 1991 and many East Timorese were injured.

References

1. The figure of 200,000 was both verified and qualified by David Bull, Director of Amnesty International UK, in a letter to *The Times*, London, on 6 August 1996:

   In your leading article of 1 August you cite Amnesty International's estimate of the number of East Timorese killed by the Indonesian government since 1975 as 200,000. This figure does not relate only to East Timorese killed by the security forces; it also includes those who have died of starvation or disease since Indonesia invaded in 1975. It represents a third of the population of East Timor.

   This clarification does not detract, of course, from the depressing human rights situation in Indonesia as a whole: there have been human rights violation on a staggering scale . . . thousands of citizens have been killed, political and criminal prisoners have been routinely tortured, and thousands of people have been imprisoned following show trials, solely for their peaceful political or religious views. The violations are continuing and unless concerted domestic and international pressure is applied on Jakarta there can be little prospect of real improvement.

2. Writing in the *Sydney Morning Herald*, 21 September 1996 Max Stahl said:

   In the war of symbols the mere survival of the Timorese guerrilla force is the most potent threat to Indonesia's ruthless campaign to integrate the former colony into their empire . . . Indonesian claims of mass resistance surrender, like pictures of 'former guerrillas' swearing loyalty in bizarre ceremonies, mixing their blood with that of chickens, were exotic fabrications, Commander Santana said. Almost no-one surrendered at all.

3. The *Far Eastern Economic Review* of 18 October 1996 published a poll of business executives across Asia concerning East Timor and Bishop Belo in which 64 per cent of respondents believed that Jakarta should allow East Timor to secede from the Republic. Those supporting secession included 82 per cent of executives polled in the Philippines, 73 per cent in Japan, 67 per cent in Singapore and 53 per cent in Malaysia, which is another Islamic state. Fully 75 per cent believed that the Nobel committee was justified in awarding the Nobel Peace Prize to Bishop Belo and Ramos-Horta.

14

Guatemala: The End of a Long War

Successive editions of *War Annual* described the civil war in Guatemala between the military government, with its strong Fascist elements, and the Guatemalan National Revolutionary Unity (URNG is the Spanish acronym). The war began in 1954 but the level of armed conflict appreciably decreased in 1994 and 1995. Progress toward real peace was set back in October 1995 when an army patrol stormed into the village of Aurora 8 de Octobre where residents were celebrating the first anniversary of their return from refugee camps in Mexico.

The defence minister resigned in disgrace, President Ramiro de Leon Cærpio dismissed the regional army commander and the soldiers concerned were court martialled. According to the report, the patrol leader was executed. Skirmishes between guerrillas and the army continued but peace negotiations went ahead. On taking office in January 1996, President Alvaro Arzu Irigoyen made real progress in addressing the root causes of his country's difficulties. He committed his government to addressing long-standing grievances about health, education, social security, housing and labour conditions. He purged the military, placed it under civilian control – an astounding step – and convinced it of the need to cut its numbers and its budget. Nobody could question his energy and courage; nevertheless during 1996 the military carried out more than 100 death-squad killings.

The six years of negotiations produced a most hopeful result on 8 December 1996, when the 36-year war officially ended. More than 100,000 people had been killed and another 40,000 had 'disappeared' – which means they were probably also killed. At the National Palace, representatives from more than 40 countries witnessed four senior rebel leaders, government representatives and the UN Secretary General, Boutros Boutros-Ghali sign the agreement.

Human rights activists complained that the section of the accords dealing with pardoning combatants was so ambiguous that it was effectively a 'blanket' amnesty that forgave everybody. However, the government said the agreement had one of the toughest amnesty accords in Central America. It was so worded that people who had committed genocide, torture and other violations of international human rights conventions could be prosecuted.

Guatemala has joined El Salvador and Nicaragua in ending the wars that had destroyed much of the social cohesion and economy of much of Central America in the period 1960–80. To a large extent, these conflicts were part of the cold war and successive American governments had supplied billions of dollars in military aid – and specialist instructors – to governments that the US believed were fighting Marxist rebellions. The war in Guatemala was an American 'invention': the CIA sponsored a coup in 1954 that overthrew a left-wing government – but it had been popularly and fairly elected. The dreadful war that ensued left 80,000 widows and 20,000 orphans and one million people were displaced.

American military advisors suggested a scorched earth campaign to wipe out communities suspected of being sympathetic to the guerrillas. Innumerable atrocities were carried out by the army of government-organised 'civil defence' forces in the early 1980s. Most of the victims were poverty-stricken, disenfranchised Indians who make up 60 per cent of Guatemala's 11 million population.

I have mentioned the six years of tortuous negotiations that preceded the peace, but to a large extent the war's end was the result of war weariness. The age of military dictators was passing and civilian leaders were developing the courage to stand up to the bullying generals still in authority. Also, new, younger officers were rising to senior ranks and they were more sophisticated and aware of the need for their country to be rescued from the ruin inflicted upon it. At the same time, left-wing rebels became aware that they could no longer rely on the support of sympathetic regimes in the name of revolutionary struggle, such as those of the cold war Eastern Europe and bankrupt Cuba. Old habits die hard and peace in Guatemala could stumble into ambushes but it is a relief to be able to relegate the Guatemalan war to the category of 'wars apparently concluded'.

15

India and Pakistan

The fortunes or misfortunes of these rival countries – Hindu India and Muslim Pakistan – are inextricably linked and generally in a violent way. There are several sets of troubles – on the border between the two nations, within Kashmir and within each of the two countries.

Kashmir is a particular difficulty and it is as insoluble as the conflict in Northern Ireland, in Lebanon and in Sri Lanka. The mountainous region of Kashmir lies between the two countries' northern extremities. Since partition in 1947 India has held two-thirds of it and Pakistan one third. Successive Pakistani rulers and governments have claimed that the whole of Kashmir is Pakistan's. Conflict is continual and UN military observers, who have been there since 1949, are powerless to stop it. The war on the Siachen Glacier at 20,000 feet is probably the world's most absurd.[1]

Muslim Kashmiris demonstrate, fight, murder and take hostages in bloody attempts to wrest Kashmir from India. The Indian army regards the Muslim militants as vermin and tries to exterminate them. More than 40 militant groups are active, some of them Indian, others Pakistani; yet others want total independence for Kashmir. Foreign governments are deeply involved – Pakistan and Saudi Arabia arm the Muslim fundamentalists.

Kashmiris are Pakistani pawns. It is just not true – as Pakistan would have it believed – that Kashmiri Muslims, with their unique Sufi traditions, want to join Pakistan from a sense of Islamic brotherhood. But this falsehood has been used to justify 50 years of cross-border bloodshed. The truth is that while the Kashmiris do not like India they dislike Pakistan even more.

The Indian government can list several reasons why it must hold Kashmir and one of them is, paradoxically, to protect the lives of 100 million Muslims living in India. The Indian rulers subscribe to Mahatma Gandhi's belief that 1,100 million of numerous faiths can only be governed through secular democracy. If Kashmir were allowed to break away over religious differences a bloody Hindu backlash would take place against the Muslims living in India.

In 1994–95 many Pakistani-trained militants and Afghan war veterans infiltrated northern India to boost Kashmiri resistance to Indian rule. One of the consequences was that 200 of these men were burnt to death when Indian troops set ablaze the town of Charar-e-Sharief where they were hiding. In the unending tit-for-tat conflict, Pakistani protectors burnt down Hindu temples. They also kidnapped five Westerners on a walking holiday in Kashmir; they killed one of them and nearly two years later were still holding the others, if indeed they were alive.

Pakistan's Arms Build-Up

Pakistan is aided by China and in September 1995 the US Senate cancelled a five-year ban on military aid for Pakistan and authorised the sale of weapons worth $370 million. The arms package was considerable. It included three P-3C Orion maritime patrol aircraft, engine upgrades for P-16 fighters, night-fighting equipment for helicopter gunships, radar systems and several types of missile including 28 Harpoon anti-ship missiles. Simultaneously, Pakistan had bought three Agosta 90-B Diesel-electric submarines armed with Exocet missiles and was negotiating to buy 40 Mirage 2000-B fighters.

Unsurprisingly, India was alarmed about Pakistan's growing strength. When they heard that Pakistan was negotiating with Ukraine to buy up to 350 T-738 main battle tanks, 152mm self-propelled guns, APCs, air defence systems and MiG-29 fighters their alarm turned to desperate anxiety.

In July 1996 Pakistan did buy about 300 T-738 main battle tanks from Ukraine. This announcement came from Defence Minister Aftab Shaaban Mirani, who would have preferred T-72M1 tanks from Poland. Influential Pakistani generals forced Mirani to opt for the Ukrainian product, which is built by the firm Khar'kov. The tank is armed with the 2A46M-1 smoothbore 125mm main gun and can also fire the 125mm 9K120 Svir laser-guided missile, generally known as the 'AT-11 Sniper'.

During the Cold War much of India's defence needs had come from the Soviet Union, a close ally. But with the disintegration of the Soviet Union, the Commonwealth of Independent States (CIS) needed hard cash and have sold Soviet-made equipment to the first buyer, giving preference, however, to Islamic states, such as Pakistan.

The Americans took an unusual step in order to show India that their arms sales to Pakistan did not mean that they favoured that country over India: in 1996 they organised their first joint naval manoeuvres with India in the Indian Ocean. Senior military leaders also made reciprocal visits to each other's countries.

Other US attempts to pacify India did not work. For instance, America pressured India to defer indefinitely the mass production and introduction of its Prithvi, a short-range surface-to-surface missile. In the face of Pakistan's heavy build-up, India ignored the American request and during 1996 and into 1997 Prithvi missiles were being delivered to the army, with a more sophisticated version promised to the air force.

Indian Army – Weak in Officers, Strong in Artillery

India has the world's third largest army with 1.2 million soldiers but its chiefs feel themselves to be at a qualitative disadvantage compared to Pakistan. Pakistani intelligence was delighted to read in an Indian parliamentary report on defence manpower and management that the army was short of about 15,000 officers, mostly captains and majors. The report was 'restricted' rather than 'secret' and it provided interesting information.

The shortage of officers was having an adverse effect on the performance of units in Kashmir and the north-east, the report stated. Combat units were operating with fewer than half the officers required, while the navy was 800 officers short and the air force 1,000. Moreover, appeals to the government to attract talented people into the forces had been ignored.

In the Indian army the average age of a colonel commanding a regiment is 46 but this is considered to be too old for the modern battlefield. This statement in the parliamentary report confirmed the army's own assessment after study of its performance in Sri Lanka in 1988. While there were too few junior officers there were too many very senior ones. The report quoted instances of lieutenant generals carrying out tasks normally handled by lieutenant colonels.

Pakistani analysts of the Indian report were pleased to find that training establishments were not being properly used. They included the National Defence Academy, the Indian Military Academy and the Officers' Training Academy at Madras. Personnel coming out of these establishments were professionally inferior to those of a decade earlier.

Apparently, the main criticisms of an army career are low wages, loss of prestige in the public's estimation, and too little leave. Many officers take early retirements when they find promotion avenues blocked beyond the rank of major.[2]

Pakistanis have a paranoid fear of India. The truth is that India has neither territorial designs nor aggressive intent. Its politicians, unless goaded, use moderate language towards Pakistan. India has offered to pull back from Siachen but Pakistan replied in an offensive way. India has offered to reopen crossborder trade but well into 1997 Pakistan had not replied. India suggests bilateral talks on Kashmir but Pakistan declines. Delhi is willing to declare the boundary in Kashmir an international border but Pakistan will not agree, though it must be evident to its leaders that there is no prospect of India ever abandoning the Kashmir Valley, which is a buffer against Chinese expansionism.

Why is the Pakistani establishment determinedly waging a Cold War against India? Cynical realists say that without enmity against India the reasons for Pakistan's existence would be questioned.[3]

Another matter relating to India should exercise Pakistan's leaders more than the army's shortage of officers – India's increase in artillery. The army's first artillery division became operational in December 1996. The use of guns is replacing the old doctrine of manoeuvre, so that artillery is no longer a support arm but a combat arm in its own right. The thinkers in the Indian army say that deep thrusts with mechanised forces no longer make sense, especially with the introduction of ballistic missiles into the Indian artillery. What matters, they say, is for any potential enemy to realise that India has enough artillery to impose a counter-attack of attrition. Any enemy columns of armour and artillery would simply be annihilated. Yet another factor is involved: Pakistan has developed strong concrete fortifications along the long frontier. These and supporting field works could only be crushed with heavy shells.

The Indian army uses artillery of 14 different calibres to cover all the potential war situations. One might be an independent guns-only operation, another close support of other arms in a tactical battle; a third role – and this is seen as most likely – is counter-bombardment. The army seems to have solved the logistical problems of supplying its many artillery units using so many different types of ammunition. The bulk of the artillery force is its 195 field regiments. The field artillery's longest range weapon is the Prithvi SS-150 surface-to-surface missile, and No 333 Missile Group was formed in 1995 and based at Secunderbad to man the missile-launchers.

A Hunger for Nuclear Capability

Pakistan will stop at nothing to gain nuclear technology. During the 1990s it recruited a covert network of scientists studying nuclear disciplines at British universities. Their mission was to gather technology for the 'great bomb project'. The people concerned were all British citizens of Pakistani origin in a position to undertake 'sensitive' research.[4]

Much of the information concerning this plot came to light after a deportation order was served on Muhammad Saleem, 51, a clerical officer at the Pakistan High Commission in London. MI5 investigators found that Saleem, who had been in Britain for 21 years, was the head of the spy network. The British Home Office, to which MI5 reports, disclosed that he had worked as secret representative for Khan Research Laboratories (KRL), Pakistan's nuclear research institute. In a statement to the court, the Home Office said:

> KRL is involved in the research, development and covert procurement of sensitive nuclear equipment which can have a nuclear weapons application. Since 1990 Muhammad Saleem has been conducting covert nuclear procurement activities in Britain on behalf of KRL.

While the Pakistan government denied that KRL was involved in military nuclear programmes, it had no response to the disclosure that KRL had bought from China 5,000 ring magnets, which are used to help enrich uranium in nuclear weapons. Saleem's appeal against deportation was dismissed but India, which saw itself as the 'victim of Pakistani nuclear aggression', protested that KRL and Pakistan generally had been treated too leniently for too long in its conspiracy in Britain.

While Pakistani troops have formed part of UN peacekeeping forces, the country itself is not exactly a role model for peace. Many Islamic fighters went to Pakistan to take part in the *jihad* against the Soviet Union in Afghanistan. After the Russian withdrawal and the fall of the communist government of Afghanistan in 1992 some of these men left Pakistan but many, being wanted terrorists in their own countries stayed behind. Pakistan not only gives them sanctuary but permits them to export terrorism. Some of Egypt's worst terrorists live in settlements along the Pakistan-Afghanistan border. Both *Al Gamaa al-Islamiya* and *Al Jihad*, violently extreme groups, claimed responsibility for the bomb in the Egyptian Embassy in Islamabad which killed or injured nearly 100 people.

MINI-WAR IN ASSAM

One of many mini-wars going on within India's borders or close to them concerns the United Liberation Front of Assam (ULFA) which has waged a secessionist campaign in the Indian state of Assam since 1979. In February 1996 Indian troops crossed into the Himalayan kingdom of Bhutan and attacked a ULFA training camp. The raid was pre-planned and not a 'hot pursuit' operation but the Bhutanese authorities made no public protest about violation of their borders.

The ULFA is led by Arabindra Rajkhowa, who has several other bases in Bhutan. The ULFA fighters trained there launch their attacks from the bases and find refuge in Bhutan later. Other ULFA camps are in the Sylhet and Khagrachari regions of Bangladesh. One of the few journalists who has visited the camps says that each is garrisoned by up to 50 guerrillas.

The Indian Defence Ministry claims that much of the ULFA's equipment and weapons is fed to the organisation through Thailand, which has a thriving arms trade. Bhutan's government does not actively support the ULFA insurgency but neither does it do much to crush it. Officials in the Bhutan capital, Thimpu, say, in effect, 'Our country has warm relations with India but we do not have the forces to engage the ULFA'.

Casualties in this small war cannot be determined but that they irritate the Indian government is clear for a source in New Delhi which says that the army would like to use overwhelming force to crush the ULFA but fears an international outcry.

The Mohajir Quami Movement (MQM) in 1996 launched an armed campaign for greater autonomy. The MQM activists claimed that they were denied government jobs and had been reduced to a minority in their own Pakistan province of Sindh.

More than 15,000 people have been killed in Kashmir's civil war; another 2,000 have fled in the MQM's campaign of 1995–97.

INDIA REVIVES ITS MISSILE PROGRAMME

Building an entire anti-missile defence system is an ambitious programme but India announced such a move in August 1997. The system is based on the Agni (meaning fire) intermediate-range ballistic missile. The Defence Minister, Mulayam Singh Yadav, revealed to parliament that 'India has no option but to continue to develop its missile capabilities against the adventurist intentions of a hostile country'. This was a reference to not one but two hostile countries, Pakistan and China. The two are allies.

After its third test in February 1994, it reached a range of 1,400km. It is known that eight more tests are needed to bring Agni to its optimum range of 2,400km. The further upgrading of Agni was triggered by Pakistan's testing of its Hatf-111 missile in June 1997. While it has a range of only 800km it could hit major Indian cities. Defence Minister Yadav announced that the Prithvi (meaning earth) indigenously-developed surface-to-surface missile had gone into full-scale production. The Prithvi was no secret and has been mentioned in the two previous editions of *War Annual* but it was not then in production.

Taken together, the Agni and Prithvi illustrate India's doctrine of deterrence; at least, this is how India's government wants them to be seen. Its spoksmen repeatedly stress that India has no hostile intent with its development of vastly expensive and sophisticated missiles, but Pakistan is not convinced and just as repeatedly warns its ally China that Pakistan is at risk. My judgment is that India is at more risk from Pakistan.

References

1. I have described the 'progress' of this extraordinary and pointless conflict in earlier editions of *War Annual.*
2. Elements of the Indian parliamentary reports were published in *Jane's Military Exercise & Training Monitor*, July–September 1996.
3. See 'Fear and pointless wars tear Pakistan apart', by Christopher Thomas, *The Times*, London, 25 September 1996. 'The military has ultimately held Pakistan together and the army is still the most powerful force for stability – and the only one not disintegrating. The courts are corrupt, the police venal, the politicians rotten. This is why Pakistan has been under military rule half its life.'
4. Reports of Court proceedings.

16

Iran: The 5-Star Danger

The term 'flashpoint' has become commonplace in assessments of dangers posed by regimes which base their foreign policies and strategies on terrorism and violence. A cursory computer count of reports in the print media would show that Algeria, Libya, North Korea, Cyprus, Iran, Iraq and Afghanistan and the insignificant Spratly Islands (South China Sea) were labelled flashpoints in 1996. Of all these countries, Iran is probably the 5-star flashpoint, for these reasons:

- It is still driven by the fundamentalist forces unleashed by the late Ayatollah Khomeini.
- The 'Iranian Revolution' – from the autocratic rule of the late Shah to the religious rule of the mullahs – has not yet run its course.
- The hatred of the rulers for the Great Satan (the United States) and the West generally is undiminished (though many of the Iranian people have recovered from this sickness).
- The eight-year war against Iraq left Iran embittered against all Arab states. (Iran is a Persian, not an Arab, state.)
- Iran seeks to destabilise certain Gulf regimes, notably Bahrein.
- Iran owns three strategic islands, Abu Musa and Greater and Lesser Tunba, in the chokepoint of the Strait of Hormuz and will fight against any threat or perceived threat to these islands.
- Iran is the leader of the incessant campaign to eradicate Israel; nothing less will satisfy the ayatollahs.
- The leader of world terrorism, Iran assassinates its opponents in the West.
- Iran is one of the world's leaders in the experimentation and development of biological weapons. Like Iraq, Iran has used such weapons and in some ways is ahead of Iraq in the ability to deliver them. Difficult though it may be to believe, Iran would more readily poison water supplies than would Saddam Hussein.

Above all these factors of war is the reciprocal hostility between the US and Iran. Each frequently accuses the other of arrogance and terrorism and their war of words often reaches the threat level. At times during 1996–97 it seemed that open confrontation might take place.

The presence of American servicemen in the Gulf is itself a cause of tension, perhaps the greatest tension. US ships patrolling close to Iran's shoreline is offensive to the ayatollahs and when 'enemy' manoeuvres are held in the Gulf they become furiously angry. At the end of 1996 the US, with a force of 24,000, conducted Operation Rugged Nautilus, a 60-day series of 'exercises' from bases in Bahrain, Saudi Arabia and Qatar.

This was part of the Clinton administration's Dual Containment Policy (DCP), the declared aims of which are to use economic and military pressure to radically modify Iranian political behaviour and totally destroy Saddam Hussein's regime. For the Americans, DCP must achieve protection of Israel in the region and ensure the safe passage of oil through the Strait of Hormuz.

Iran – Bahrein

Bahrein's relatively poor Shia minority calls for the enforcement of strict Islamic law and for economic reform. The Bahreini Shias are manipulated from Iran, the home of Shia fundamentalism. The Teheran Shia administration of President Ayatollah Rafsanjani has long wanted to end the Bahreini ruling al-Khalifa family's close ties with Washington, London and Paris.

In January 1996 the US exchanged large quantities of surplus military equipment for basing facilities in Bahrein. The al-Khalifas want US forces permanently in Bahrein not only as a protection against Iran but against neighbouring Qatar; Bahrein and Qatar are in dispute over the Hawar Islands and fishing rights.

Opposition to the Bahreini government extends beyond the island itself and its neighbours and is significant in its influence. The London-based Bahrein Liberation Movement alone claims to have 50,000 members. The al-Khalifa family is worried about any dissent following the palace coup in Qatar in 1995 and has radically increased opposition to the governments of Saudi Arabia, Oman and Kuwait.

In June 1996 the government foiled an Iranian-backed coup attempt and arrested 30 plotters, some of whom had been trained by Islamic Revolutionary Guards in Iran and Lebanon. The coup attempt had serious implications because of the danger of any unrest spreading along the causeway which links Bahrein to the Shia-populated eastern province of Saudi Arabia and from there throughout the Gulf. The Saudis give Bahrein extensive military support. The Bahreini Minister of Information, Muhammad Ibrahim al-Mutawae, announced at the time of the plotters' arrest:

> This is a serious conspiracy which reveals that an organisation known as the military wing of *Hezbollah*-Bahrein, together with the Iranian authorities, has been planning since 1993 to undermine Bahrein's security and stability. The movement's main aims are to overthrow the Bahrein govern-

ment by force and to replace it with a pro-Iranian regime.

Western intelligence knows that *Hezbollah*-Bahrein was founded in 1993 in the Iranian holy city of Qom. During 1996 its terrorists made bomb attacks against luxury hotels and opulent shopping areas as well as state-owned targets. In the filthy slums of the Shia people black flags hang over many a house in memory of the 'martyrs' who fell in riots against the authorities.

The United Arab Emirates (UAE) also accuses Iran of developing an 'offensive rearmament programme and building up its forces and equipment on Abu Musa and the Greater and Lesser Tunbs'. According to UAE Foreign Minister Rashid Abdullah:

> Iran's weapons there are not defensive. They are offensive and are directed against the Emirates and other countries in the region. There is a strategic imbalance in the region, especially concerning Iran and Iraq.

Warnings to the Americans

In May–June 1996 Iran staged Operation *Valayet* (Guardianship) in the deserts south of Teheran. More than 200,000 soldiers and airmen took part in the manoeuvres, the largest peacetime operation since 1979. It was intended to demonstrate Iran's military strength and readiness and it gave the Land Forces Commander, General Ahmad Dadbin, an opportunity to warn the US. 'We are strong enough to defend ourselves, whoever the aggressor may be', he said, 'and the Americans should think twice before launching any attack.'

The US had at that time issued no particular threat of military action, despite repeated statements that Iran was guilty of sponsoring terrorism and that it was seeking to acquire nuclear weapons. It is certainly true, as General Dadbin admitted, that Iran has achieved 'the technology of missile production'.

Iran has nuclear research facilities, including what is supposed to be a secret plant at Neka, on the Caspian coast, north-east of Teheran. In mid-1996 Israeli defence chiefs made thinly-veiled threats to knock out this plant. In Teheran, any Israeli threat is perceived to be US-inspired. In May 1996 Iran test-fired a new missile, the Tondar, a shore-based anti-ship missile that is made in Iran.

Iran's main nuclear weapons facility is 25 miles north of Teheran in the Argoz Mountains and is one of several 'contingency targets' for the US, should Iran seek to take over the Middle East, as the CIA believes it intends to do. Other targets are the CSS-2 surface-to-surface missiles guarded in tunnels overlooking the Strait of Hormuz, through which half the world's supply of oil is exported. The CIA has located 13 terrorist training camps in Iran and these, too, would be American targets. Between 1988 and 1996 Iran spent £26 million in rebuilding its military forces – and Western intelligence knows precisely where the money has gone.

While Iran could not possibly win a war against the US alone, let alone one against a Western alliance, it would be destructive in defeat. Even if this vengeful and fanatical regime perceived itself to be under serious threat it could embark on a pre-emptive war using biological weapons against which its enemies would have little defence.

The Iranians keep stocks of toxins inducing anthrax and botulism in Tabriz, north-west of Teheran, and they could quickly build up their stocks. They could already deliver their toxins and live organisms by Scud or from Sukhoi aircraft and by the end of the decade they will have long-range ballistic missiles.

Even more frightening, the Iranians have a biological pollutant for usc in poisoning water supplies. Certain intelligence agencies are convinced that Ayatollah Khamanei, Iran's terrorism ideologue and decision-maker, would not hesitate to poison public water supplies in Europe and North America. According to one source, given credibility by the CIA and Britain's SIS, Iran's biological weapons stocks include an aerosol no larger than a hairspray or shaving cream can. Its potential as a terrorist weapon is obvious.[1]

On 21 November 1996, *The Washington Times* reported that China had sold the Iranians missile technology, components, an advanced radar system and 400 tonnes of chemicals for use in producing nerve agents.[2]

Assassination as a form of Warfare

The late Ayatollah Khomeini, who initiated Iran's Islamic Revolution, always understood that assassination was a political weapon. He came from the region where the original assassins were active in the 12th century. In preaching to his ayatollahs and mullahs as well as to the armed forces of Iran, especially the Islamic Revolutionary Guards, he made clear that in *jihad* (holy war) assassination was not only permissible but much to be desired, since it weakened the will of opponents to resist. Assassination did not mean only the murder of a single opponent, though this has generally been the understanding in the West. For Khomeini, groups could be assassinated.

Few people in the West realise that for the Iranian regime assassination is 'normal' and has an agenda. The intention since 1979, the year of the Shah's overthrow, has been to eliminate every single important Iranian dissident. In Khomeini's thinking – and later that of his successors, President Rafsanjani and Ayatollah Khamanei, Iran's spiritual leader – assassinations can be pre-emptive: they can nip uprisings, revolutions and even civil war in the bud.

Between 1979 and the end of July 1996, 215 overseas attacks had killed or wounded more than 350 critics of the regime in 21 countries. During 1996 the dead included: Zahra Rajabi, a member of the dissident Iranian National Council of Resistance, killed in Turkey, where she was investigating the plight of Iranian refugees; Reza Mazlouman, a former minister of the Shah's government murdered at home in Paris; two Sunni clergymen killed in Pakistan; and four Iranian Kurds who had fled to Iraq.

According to the US State Department, the Iranian paymasters hand out $100 million annually to Islamic groups to finance terrorist operations. In June 1996, the British Parliamentary Human Rights Group produced evidence that Iranian assassins had travelled abroad to kill 11 critics of the regime in the first five months of the year, more than in the whole of 1995. Of course, these are only the *known* assassinations. In its report the Parliamentary Human Rights group stated, 'The use of terror as an adjunct to foreign policy has developed into an organised and professional activity'.[3]

In March 1996, Belgian customs officials in Antwerp checked the Iranian freighter *Kohladooz*. They discovered an advanced mortar-launcher and shells; the mortar has a range of 700 metres and the shells were supplied with proximity fuses so that they would detonate above their targets. According to Belgian and French anti-terrorist police, the most likely target was the Paris HQ of the Iranian resistance group and its national president-elect (should the ayatollahs' regime be overthrown), Maryam Rajavi.

On 25 June 1996 a huge bomb devastated the US barracks in Dhahran, Saudi Arabia, killing 19 American airmen and injuring scores more. While there was no firm evidence that the bomb had been planted by Iranian agents much circumstantial evidence pointed in that direction.

Early in 1997 Iraqi terrorists launched a mortar on an Iranian opposition group in Baghdad, Iraq, killing several people and wounding scores more. The National Council of Resistance of Iran, which maintains a dossier on the 'super-mortar', reported that the order to increase production of the weapon came from Ayatollah Khamanei. That this claim was not simply a propaganda blast from the regime's opponents was borne out by European intelligence agencies which reported, in January 1997, that Iran had provided 15 mortars to terrorist cells throughout Europe.

Britain's MI6 reported to the government that Teheran was considering supplying the IRA with a version of the mortar, 'made to order' by Iran's Organisation of Military Industry. Iran has given the IRA funds in the past and the ayatollahs have based one of their largest intelligence operations in Dublin, from where they are in direct contact with the IRA.

It should be noted that the population of Iran has doubled to 60 million since the revolution and the country's economy is in trouble, with 60 per cent inflation. Large numbers of the poor survive on bread coupons from the state. Repression is everywhere and the ayatollahs' stringent rules about behaviour are enforced by 30,000 'guardians of the revolution', who act with great brutality. An underground pro-democracy movement exists but it is a long way from toppling the ruling clerics.

Refsanjani's Successor

In May 1997 the Iranians went to the polls to vote for a President for the next four years. The result, a landslide of 69 per cent for Muhammad Khatame, surprised most observers since Khatame was known to be a relative moderate. Compared with Rafsanjani and many other ayatollahs he is certainly a moderate. Foreign analysts attribute his victory to the votes of the young, women, technocrats and the Left.

Khatame assumed presidential power at the beginning of August but in practice his power has limitations. While he is head of the executive and appoints ministers, parliament must approve his selections. Supreme authority rests with the Spiritual Leader, Ayatollah Khamanei, whose extreme foreign policy views are well known. They were obviously well known to President Clinton who, at the time of Khatame's political victory, called it no more than 'hopeful'.

Nevertheless, in one of his first policy statements Khatame appeared to signal a change in Iran–US relations when he did not use the hostile rhetoric of the past. His key sentence: 'Any changes in our policies towards the USA depend on changes in the attitude and positions of the USA concerning Iran's Islamic revolution'.

It was significant that several regional governments, notably Bahrein and Egypt, welcomed Khatame as president and wished him success. Everything depends on what his opponents permit him to do. The Speaker of the Iranian parliament, Ayatollah Ali Akbar Nateq-Nouri, who was expected to win the presidential election, holds strong anti-West views and during the campaign criticised Khatame for his 'softness and lack of resolution'.

Khatame's election was not enough in itself to have Iran downgraded from being the 5-star danger.

References

1. Michael Eisenstadt, senior research fellow at the Washington Institute for Near East Policy, testifying at the meeting of the Committee on Foreign Relations of the US Congress, 19 March 1996: 'Terror and subversion have been key instruments of Teheran's foreign policy since the Islamic Revolution in 1979. Since then Iranian-sponsored and inspired terror has claimed more than 1,000 lives worldwide.'
2. The sources for information about Iran's terrorist agenda and stocks include: a Mossad assessment; CIA report sent to the US Senate Intelligence Committee, 31 July 1996; another CIA report entitled 'Arms Transfers and State Sponsors of Terrorism', dated 2 October 1996. This report was labelled 'top secret' and was apparently for perusal only by senior military and political leaders.
3. The report states: The operations of the state terrorists have not been limited to Iranian citizens. Several of the translators and publishers of Salman Rushdie's *The Satanic Verses* have been attacked. The Japanese translator was murdered in 1991; the Italian translator escaped an attempt on his life as did the Norwegian and Turkish translators in 1993. The *fatwa* against Rushdie was reiterated by President Rafsanjani himself on 3 February 1993 and the reward offered for his murder increased to $2 million. The Speaker of the Iranian parliament, Ali Akbar Nateq-Nouri, said on the fifth anniversary of the *fatwa* that 'every Muslim is religiously bound to kill Rushdie whenever and wherever he can be found'.

17

Iraq, the US and the Kurdistan Conflict

DESERT STRIKE – Saddam Hussein vs Bill Clinton

The 'war' in Iraq in 1996 was actually several conjoined conflicts, involving Iraq, Iran, two opposing Kurdish factions, the US and Turkey. Numerous other countries, including Britain, were marginally involved. Overall, the political-military-intelligence situation was complex.

The Gulf War which followed Saddam Hussein's invasion of Iraq took place in 1991 but it never really ended because it was necessary for the Western Allies to ensure that the ever-aggressive Saddam kept to the UN's end-of-war terms. A selected list of 'incidents' show that tension was continual.

1991:

20 March: A US F-15 shot down a Su-20 when the pilot defied an allied ban on military flights.

22 March: US F-15 shot down a Su-22 while it was attacking Kurdish rebels.

1992:

27 December: An F-15 shot down a Mig-25 in the southern no-fly zone.

1993:

13 January: Concerned about the danger posed by SAM sites in southern Iraq, the American, French and British allies sent 116 aircraft in a night raid against the sites, destroying four of them.

17 January: Allied planes attacked a missile site in northern Iraq; an American F-16 shot down a Mig-23; US warships fired about 50 cruise missiles at nuclear plants.

18 January: Allied aircraft attacked the SAM sites which had not been damaged on 13 January; 80 British, American and French aircraft took part in the raid, the first major day assault. Simultaneously American and British planes bombed missile batteries in the north.

19 January: The Allies came under fire and responded with attacks against Iraqi defences.

27 June: Following an abortive attack to assassinate the then President Bush, US warships in the Gulf and the Red Sea fired 13 Tomahawk missiles at Saddam's intelligence HQ in Baghdad.

29 June: Targeted by Iraqi radar, a US F-4G fired a missile at an Iraqi anti-aircraft battery.
25 July: In a similar incident, another F-4G fired at a missile site.
29 July: US Navy EA-6B aircraft fired at missile sites.
19 August: Anti-aircraft batteries fired on two American F-16s, which then dropped cluster bombs on the gun-sites. Soon after, laser-guided missiles were fired at other Iraqi weapon sites.
1994:
October: Saddam marched 70,000 troops to the Kuwaiti border and the US carried out a major deployment.

Kurdish Tribal Hatreds

The 1996 wars in Iraq were largely the result of the intense reciprocal hatreds between the Patriotic Union of Kurdistan (PUK) led by Jalal Talabani and the Kurdish Democratic Party (KDP) of Massoud Barzani. Without an understanding of their animosity and political objectives it is not possible to follow the Kurdish conflict, which is intricate enough without the devious involvement of the Iraqis, Iranians, Turks and the Western allies of the Gulf War of 1991, notably the US and Britain, but *especially* the US.

As has been described in previous editions of *War Annual*, the Kurds have been plagued for centuries by disunity. More than 20 million ethnic Kurds are spread across Iraq, Iran, Turkey and Syria and that they still lack a homeland of their own is largely due to generations of rivalry, deceit and missed opportunities. They had their best chance in 1991 when Saddam unleashed an onslaught against them with his many surviving helicopter gunships. In despair, in disarray and with many casualties, the Kurds fled into the northern mountains.

When the Western allies created an autonomous safe area for them the Kurds held democratic elections and set up a capital in Erbil but nobody was surprised when the old tribal rivalries destroyed the budding democratic union.

The KDP of Massoud Brazani, whose base area is north-west Iraq, is tough and tribal and it controls the lucrative smuggling routes between Iraq and Turkey. Barzani's troops are equipped with small arms, light artillery, rocket-launchers and some surface to air missiles. Many Western observers refer to Barzani's men as guerrillas but this is apt only in that they are not equipped or organised to fight pitched or manoeuvre battles against large enemy forces. A better description for them is that they are irregulars.

The other major Kurdish faction is the PUK of Jalal Talabani. His rank and file members might be 'ordinary' men but many are urban and educated and some, like Talabani himself are intellectuals. His PUK, based in north-east Iraq, has 12,000 troops armed with tanks, mortars, anti-aircraft guns and surface-to-air missiles.

Barzani and Talabani hate each other and are totally at variance on 'self-rule'. Talabani consistently says he will accept only full independence for a Kurdistan that could have a seat at the UN. Barzani would accept autonomy within Iraq. Talabani's people are squeezed between Iran and Iraq, an unhappy and dangerous position, though Iran supports Talabani. Talabani is extremely jealous of Barzani because of the 'customs tax' which he collects from the unending line of tankers and trucks which transport oil from Iraq to Turkey and cigarettes from Turkey to Iran in violation of UN economic sanctions imposed against Iraq following the invasion of Kuwait.

Warfare between the PUK and the KDP is endemic and 4,000 Kurdish fighters lost their lives between 1991 and 1997 in savage encounters. In that period the US spent $900 million and worked assiduously to bring the Kurds together in order to form a united anti-Saddam front. Innumerable promising ceasefires have broken down.

The Battle of Hamilton Road

The 1996 wars in Iraq date from 26 July when an Iranian mechanised brigade ventured into northern Iraq, that is Iraqi Kurdistan. Teheran claimed that the PUK asked for Iranian help, though the PUK denied this. On 29 July the brigade withdrew but the KDP refused permission for it to pass through the KDP area. This frustrated Iran's objective of demonstrating to the world that it had the support of both the Kurdish factions.

On 17 August a major battle took place between the factions when the PUK attacked the KDP in what I shall call the Battle of Hamilton Road, a strategic route named after a 19th century British engineer. The PUK stated that its attack was to defend the populace against 'KDP murderers who are like Hitler's SS and are killing people indiscriminately'. UN observers placed the blame squarely on the PUK and stated that they, not the KDP, were the aggressors. On that same day, in London, many KDP exiles celebrated the 60th anniversary of the KDP's foundation.

The KDP's counter-attack at Hamilton Road killed 400 PUK fighters and routed the rest. The PUK then appealed to the Iranians for help. To cover up the defeat, Talabani announced that the PUK had not been fighting at all – the conflict witnessed by outsiders had merely been a violent dispute between opposing factions of the KDP![1]

The following day Barzani, in a panic, gave his first warning to the US State Department that the KDP would soon come under attack from Iran but he received no American response. Iran did indeed, on 21 August, directly enter the KDP–PUK conflict, mostly to try to prevent Kurdish incursions into Iran. Barzani told President Clinton that without American support he had no option but to turn to Saddam Hussein for help. This 'ultimatum' must have alarmed the Clinton administration but it did not bring instant help from the Americans.

On 23 August the PUK and KDP met in another bloody tribal battle which the KDP won after more than 400 fighting men were killed. A hastily arranged

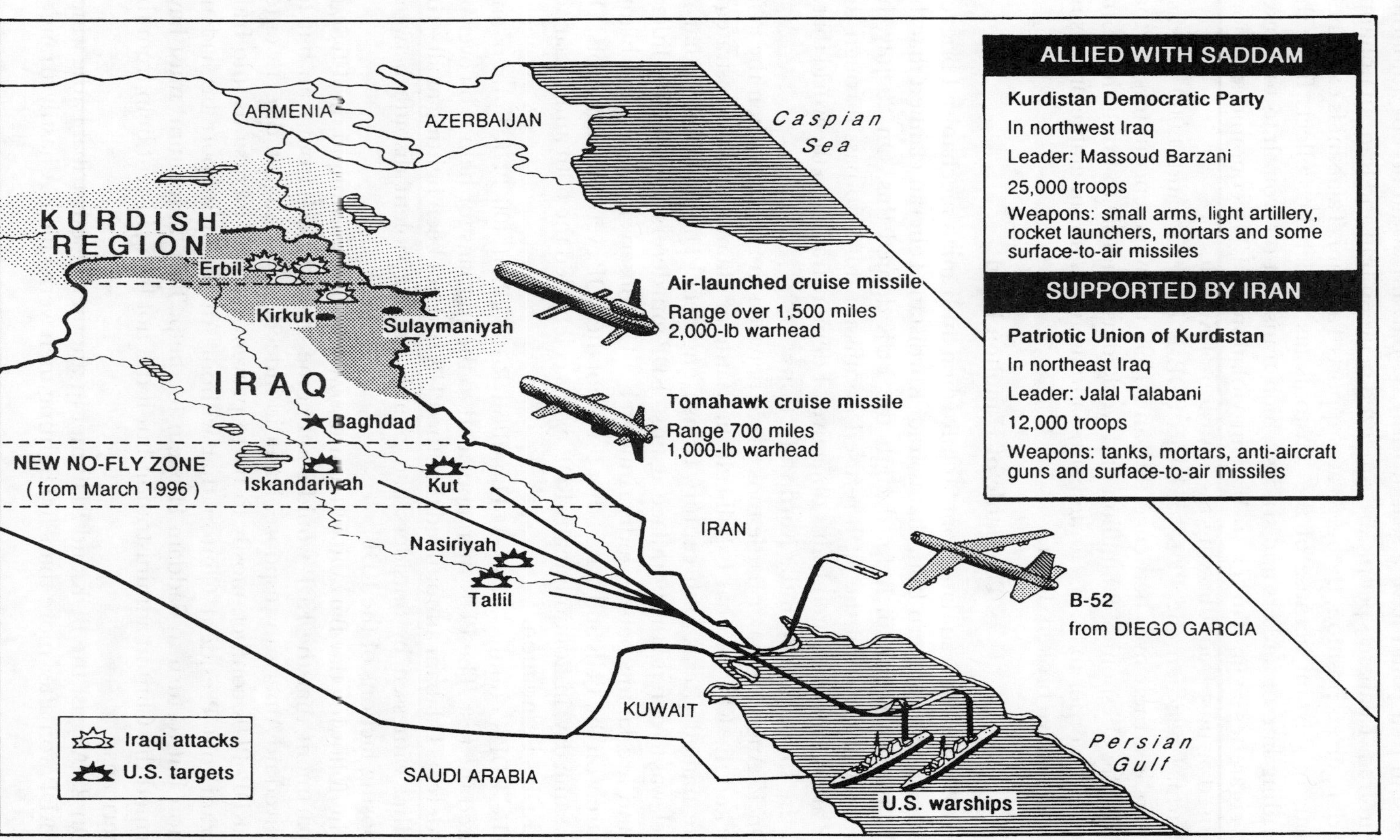

Iraq Conflict in 1996

ceasefire lasted only a short time and KDP and PUK negotiators then travelled to London for talks which, like all previous 'talks', quickly broke down.

Also on 23 August, Barzani sent an urgent message to Saddam Hussein, in fawning tones, 'inviting' him to send an army into the Kurds' 'safe haven' – which had been established by the Allies – to destroy the PUK and, as Barzani said, to prevent Iran from occupying Kurdistan. He took this step, he said later, because he still had no offer of help from the Americans. His position was even more serious.[2]

The various US intelligence agencies were assessing developments and their satellites picked up the Al-Abed division of Iran's Republican Guard, based in Kirkuk, moving towards the PUK stronghold of Erbil, not to fight the PUK but to reinforce it. American military chiefs at once ordered a state of high alert for the aircraft carrier USS *Carl Vinson* and the USS *Tarawa* amphibious group in the Gulf and the USS *Enterprise* and its carrier group in the Mediterranean. Four B-52 bombers were flown to a base in Guam, ready for forward movement. Iran backed down, with President Rafsanjani explaining that 'We only targeted our own anti-revolutionary elements'. He was referring to the rebel Iranian Kurds, who could be expected to help any coalition of forces staging a civil war against the ayatollahs' regime.

Clinton's Warning: 'A Price to Pay'

President Clinton and his advisors pondered over the developing crisis and on 28 August, when it would not resolve itself, Clinton sent a stern diplomatic *démarche* to Baghdad warning Saddam not to interfere in the safe haven. Three days later he despatched another 'stay put' instruction. Saddam Hussein remained defiant and on 31 August ordered his generals to assist the KDP to drive the PUK out of Erbil.

Clinton was preoccupied with his election campaign but he instructed his aides to induce the Kurds to cool their conflict. At the same time, he warned Saddam, yet again, that military intervention in Kurdish affairs was 'not an action you can take without paying a price'. When this was ignored, Clinton's ministers could only denounce Saddam while the President thought out his next step, for a forceful move there had to be if he was to avoid the criticism that he was too weak to stand up to the tyrant Saddam. Some of his advisors suggested a strike on Baghdad, others proposed targeting the invading Iraqi army of 41,000 men and 300 tanks.

But Clinton and other key figures in his administration saw an opportunity to gain control over a larger area of Iraq's airspace and to humiliate Saddam in front of his own military.[3] To do this Clinton would make the US strike in southern Iraq. Nobody had yet found a way of humiliating Saddam and Clinton should have known better than to try.

In military terms, the Iraqi columns' advance north towards Erbil can only be described as masterly and it owed much to Soviet/Russian doctrine. The Iraqi generals employed the *mashirovka* technique. *Mashirovka* is the Russian term for both camouflage and deception and entails driving by night, lying up under camouflage by day and timing all movements to the known patterns of US satellite observation. The Americans picked up the Iraqi movements not from satellite and air reconnaissance but only from military radio intercepts.

American retaliation against Saddam's thrust into Erbil began after the Iraqi column withdrew but before all the troops had moved out of the Kurdish region. At 4.15am GMT, on 3 September, the B-52 bombers which had been on standby in Guam but were now in the air near Iraq fired 27 missiles at fixed air defences. Two warships on station in the Gulf also fired on these southern targets.

The next day, about midnight GMT, 17 more cruise missiles were launched at four targets whose destruction in the first attack could not be confirmed. An Iraqi radar station tracked a US jet, which responded with a high-altitude anti-radiation missile (HARM). On the same day the US State Department announced a new no-fly zone, extending it 95 miles from the 32nd to the 33rd Parallel. Predictably, Saddam Hussein and the Revolutionary Command Council said that Iraq would fight the enforcement of the no-fly zones. Extension of the no-fly zone meant that about half of Iraq, including several important air and military bases, now lay within one or other of the two no-fly zones.

The American cruise missiles hit only 40 per cent of their targets. In any case this equipment was of little use against the grounded Iraqi air force. Clinton's purpose in extending the no-fly zones was to make it harder for Saddam to move aggressively southwards and threaten the strategic oilfields of Kuwait and Saudi Arabia.

Clinton never did specify the objectives of his mission against Iraq and neither he nor his principal secretaries of state were specific about its results, yet the president announced with satisfaction, 'Our mission has been achieved'.

A former Reagan National Security Council staff member, Geoffrey Kemp, said, 'We're going to be buzzing bloody Baghdad every day. It's like a psychological noose tightening around Saddam.'[4]

Saddam's Achievements

It was not clear then or later how harmed Saddam had been by Operation DESERT STRIKE. He was uncowed, he did not act like a man with a noose around his neck and he was sure enough of himself to eliminate numerous army officers accused of plotting against him. On the other hand, the US suffered some unfortunate repercussions from the operation. The UN Security Council, under Russian influence, could not agree on a resolution condemning the Iraqi attack, though it did extend the period of sanctions against Iraq for a further 60 days and suspended Baghdad's right to sell oil worth $2 billion to buy food and medicines.

Some of the Arab states, even those friendly to the US, were unhappy about the American strike and some adverse international publicity portrayed the US as the bully and Saddam as the victim. This was quite wrong but perceptions count more than reality in the Middle East.

Saddam Hussein's incursion into the Kurdish region of Iraq was perhaps more of a great raid than a war but whatever the dimensions by which we assess its scale, it was an outstanding success. Saddam was pleased about the immediate results on the ground at Erbil. His 'special forces' rounded up and killed scores of Iraqi 'defectors', blew up the broadcasting base of the Iraqi National Congress (INC) – enemies of Saddam – and brought a reign of terror to the city.

Having routed the PUK, the Iraqis handed Erbil over to their new Kurd allies, the KDP. The Iraqi commander left behind a mobile armoured battalion, which dug in six miles south-east of Erbil to deter the PUK from returning. In Erbil itself numerous spies remained as well as squads of killers to ensure that anybody who spoke out against Saddam was punish. Saddam achieved much:

- By making an alliance with a Kurdish faction, even if it were to prove only temporary, he had probably destroyed for good any united Kurdish opposition against him.
- Iran was now no longer a threat to Saddam in the north. In psychological, military and propaganda terms this was satisfying for the Iraqi regime.
- In defying the West, Saddam had shown everybody that he was still a leader who could not be ignored. His army had been victorious and had not been attacked.

The Rout of the CIA[5]

Indirectly, Saddam's biggest victory was against the CIA and therefore against the US. For five years the CIA, through six case officers on the ground in Erbil, had been running a programme to induce all the many factions of Iraqi and Kurdish dissidents to unite against Saddam. They attempted to do this through the Iraqui National Congress, an umbrella group covering 19 anti-Saddam organisations. The CIA paid $10 million a year to keep the INC in business and the organisation's importance to the Americans was obvious from the level of American diplomats who met INC officials. The US Secretary of State Warren Christopher, Vice president Al Gore and US Ambassador to the UN Madeleine Albright were among them.

The CIA men in Erbil were apparently so preoccupied with the INC operation that they had no idea that Saddam was about to invade northern Kurdistan at Massoud Barzani's invitation. This was an astonishing intelligence failure. Belatedly realising their danger, the six CIA men raced out of Erbil well before dawn on 31 August, just as Saddam's column was entering the city, and headed for the safety of Turkey. They left behind them hundreds of thousands of dollars worth of computers, scramblers and satellite phones, as well as equipment used by a TV-radio station that had been transmitting anti-Saddam propaganda for 11 hours a day.

If this had been all that was lost, the disaster would have been serious though not catastrophic. But the CIA also abandoned 1,500 members of the Iraqi National Congress to whom they had promised protection. Some of them fled to the mountain town of Salahuddin but Saddam's troops killed nearly 100 INC members and 88 of the 100 who worked for the rebel TV station. The United states' credibility and reputation for protecting its friends suffered a blow from which it may never recover.[6] Another 2,000 Kurds, working in one way or another for the Americans, were hastily evacuated to Guam for safety and 'debriefing'.

The CIA's reputation in the international intelligence community was also damaged because it seems to have been ignorant of the alliance that was forming between Barzani and Saddam. Communication between the State Department and the CIA over the Saddam–Barzani link was non-existent.

Even as American missiles were hitting their Iraqi targets on 4 September 1996, KDP and PUK units were in combat at several villages near the Iranian border and outside Erbil. This fighting continued for several days. The PUK, using weapons and ammunition freshly supplied by Iran, were taking the opportunity to use them against the KDP.

While the Iraqis, the Americans, the British, Kurds and Iranians were otherwise occupied, the Turks moved into 'Kurdistan' and occupied a new so-called security area. This gives them a better position from which to prosecute their own war against 'terrorist' Kurds. The Turks are likely to be there for a long time.

The Pentagon knew that within 10 to 15 days of the US strike against Iraq's air defence capability the sites were restored to operational capability. At Al Kut, Al Iskandariyah, Al Nasiriyah and near Basra, the missile sites were not only back in business but in communication with one another; in some places new SA-2 and SA-3 missile-launchers had been added. That Saddam could so quickly restore his air defence capability was of serious if unspoken concern to the American chiefs-of-staff.

More seriously, Operation DESERT STRIKE was to put Iraqi Kurdistan back within the influence of Baghdad after a five-year absence.

Saddam Hussein's incursion put much of the PUK's territory into the KDP's hands but they could not hold it. Late in October 1996 Talabani went on the offensive against his hated rival and regained control over all his previously held territory except Erbil. In effect, this fighting restored the position that pertained in 1993, just before Talabani captured Erbil.

It is in Barzani's power to establish at least a working relationship with Talabani: all he has to do is share the revenue he gains from charging the smugglers 'custom duty'. Talabani tells every Western journalist who reaches Zahle, his base, that the money belongs to all the Kurdish people, not to the Barzani family. Barzani will never share his spoils.

It seems very likely that the ongoing war between the PUK and KDP could become a war by proxy between Iran and Iraq. The Iranian ayatollahs would like to destroy Saddam as much as President Clinton would. Perhaps, in a strange way, this makes Iran and the US allies!

Diplomatic Fall-out From DESERT STRIKE

President Clinton pronounced Operation DESERT STRIKE to be a success but in diplomatic terms in the region it was a disaster. The main damage can be summarised as follows:

Saudi Arabia The Saudi regime supports the US policy of containing Saddam but cannot stomach attacks on fellow Arabs. Even more, the regime worries about the effect such attacks have on the dangerous anti-Western fundamentalists in the kingdom, who would like to turn Saudi Arabia into an Islamic state. The US was refused permission to attack Iraq from Saudi bases.

Jordan Jordan wants to see Saddam prevented from expansion but King Hussein is fearfully aware that many of his own people support Saddam against the West. Jordan constantly teeters on the edge of political unrest. The King refused to allow a US air force squadron to use a Jordanian base.

Turkey Before beginning DESERT STRIKE the Clinton administration informed Egypt, Jordan and Saudi Arabia – but not Turkey. Yet Turkey is a fellow member of NATO and provides bases for US, British and French air force and military units. Without Operation PROVIDE COMFORT, which is based in Turkey, the Kurdish enclave in northern Iraq could not exist. Every time the US, alone or with its allies, hits Iraq the Turkish government of Necmettin Erbankan is irritated all over again by the embargo on Iraq's oil trade, which has cost Turkey $20 billion in pipeline revenues since 1991. Erbankan would like to throw out PROVIDE COMFORT and end the boycott of Iraq. Yet the Clinton administration did not consult its Turkish allies before the attack on Saddam's air defences.

Syria The US would like to have Syria's support against Saddam and Syrian troops did play a peripheral part in the Gulf War as part of the allied force. Every time that the US makes a military move in the region the Syrian regime demands to know why Washington will not be more sympathetic to Syria's claim for meaningful negotiations with Israel over the Golan Heights. Syria supports the Kurdish Workers' Party, a breakaway faction of the KDP. The Workers' Party has 20,000 members and fights a guerrilla war against the Iranians and Turks. In Syria, the Americans won no friends with DESERT STRIKE.

Russia Saddam Hussein owes Russia many billions of dollars for armaments supplied but not yet paid for. President Yeltsin wants and needs this money but is unlikely to get it while the boycott on Iraqi oil remains. DESERT STRIKE only prolonged the boycott. Even worse, it angered the Russians because they had not been consulted; Moscow knew about it only just before it took place. National Security advisor Alexandr Lebed said 'the US spits on everybody' and in more diplomatic language added: 'We are disturbed by the American trend of using extreme and radical force against the Arab world'.

Egypt President Hosni Mubarrak, who depends on the US to the extent of $2.5 billion in aid annually was careful not to condemn the American action but he warned against it. He knew that it would further inflame his own Islamic fundamentalists who would like to see him assassinated.

France When President Clinton telephoned President Chirac the French president was strangely 'not available'. Later he said that he was displeased about the operation and he refused to help patrol the new parts of the no-fly zone, north of the 32nd parallel. France wants to pick up the billions of dollars worth of work to rebuild Iraq after the Saddam regime disappears and to this end is presenting itself as the major Western friend of the Arab people. A French foreign ministry spokesman said of DESERT STRIKE; 'The context is very different. During the Gulf War there was overt aggression by Iraq against Kuwait, a sovereign state belonging to the UN. This time, northern Iraq and Iraqi Kurdistan are in Iraq.'

References

1. Such lies are a frequent professional hazard for analysts of wars but few are so flagrant and so obvious as this one.
2. Barzani's published 'invitation' to Saddam to enter Kurdistan could have been an attempt to cover up much earlier collusion between the two men. Hazhir Teimourian, a most distinguished Kurdish Journalist based in London, in a letter to *The Times*, London, on 8 September 1996, wrote:

   I have a feeling that the decision to go over to Saddam was already taken when Massoud Barzani was asking Washington for help 'against the Iranians'. As long ago as February [1996] I heard from Kurdish exiles that Barzani had received armoured vehicles from Baghdad. The correspondence may have been designed to justify, afterwards, an act which will go down in the history of the Kurds as the worst act of treason since a Kurd showed the way to Xenophon, the Greek general, after the Kurd's friend had been tortured to death in 401 BC.

3. During 1995 US–UK air sorties had numbered 90 each day. Before Saddam's attack on Erbil they had dropped to a mere four per day and most ground patrols were being carried out by Turkish troops. The Turkish government had no objection: it needed a legitimate way of intensifying its own war against the long-suffering Kurds, especially the Kurdistan Workers' Party.
4. Reported in *Time Magazine*, 16 September 1997.
5. *Time Magazine*, reporting the affair on 23 September 1996, entitled its article 'Saddam's CIA Coup'. Nothing could have better described the Agency's abject and absurd failure.
6. Rend Rahim Francke, director of the Iraq Foundation in Washington – an anti-Saddam organisation – said: 'The CIA fled and abandoned a large number of people. All Iraqis in opposition feel extremely let down by the US.'

The Kurdish War of Independence was described at great length in *War Annual 6* and again, with many more developments, in *War Annual 7*.

18

Israel: West Bank and Gaza

THE SECOND *INTIFADA*[1]

Tunnelling work on the Haram al-Sharif, regarded as the third holiest site in the Islamic religion, was the flashpoint for the dispute which quickly became known as 'the new *Intifada*'. Palestinians feared that the reopening of the tunnel was an Israeli attempt to undermine the sanctity of Muslim Jerusalem and strike a mortal blow at their religion.

The uprising against Israel was surprising because, early in August 1996, the Israel Defence Forces (IDF) placed security forces on alert when violence broke out between *Hamas*, the militant Islamic group, and Arafat's Palestine Authority (PA). *Hamas* linked Arafat and the Israelis in a declaration of war against both. On 4 August it called on its fighters to 'strike Zionist targets in response to the crimes of the Arafat regime and its oppressive militia'. The people should launch an *intifada* against the PA, *Hamas* leaders urged.

Hamas branded Arafat a collaborator and blamed the Palestinian police for shooting dead one of its members. They had justifiable complaints against the police. In the wake of suicide bombings against Israeli targets earlier in 1996, the police had arrested about 900 Islamic militants, who were then held without charge. It was difficult for the PA and its legal system to know with what they could be charged without angering *Hamas*. Arafat ordered the release of some *Hamas* prisoners in an attempt to end the uprising.

The PA routinely rejects *Hamas's* charges that it is collaborating with Israel but on some issues the PA and Israel really do collaborate. For instance, there is a joint security committee which discusses rising crime in the West Bank. Israeli and Arab criminals were co-operating in drug dealing, car theft, forgery and counterfeiting and it was in the interests of the PA and Israel to deal with these problems together. However, such co-operation inevitably leads to charges of traitorous collaboration.

Had there been some level of collaboration before the affair of the Haram al-Sharif there might not have been a new or renewed *intifada*. The Haram is the plateau on which Islam's holy Dome of the Rock and al-Aqsa mosque are built. For Jews, the area is the Temple Mount. For generations it has been a sensitive

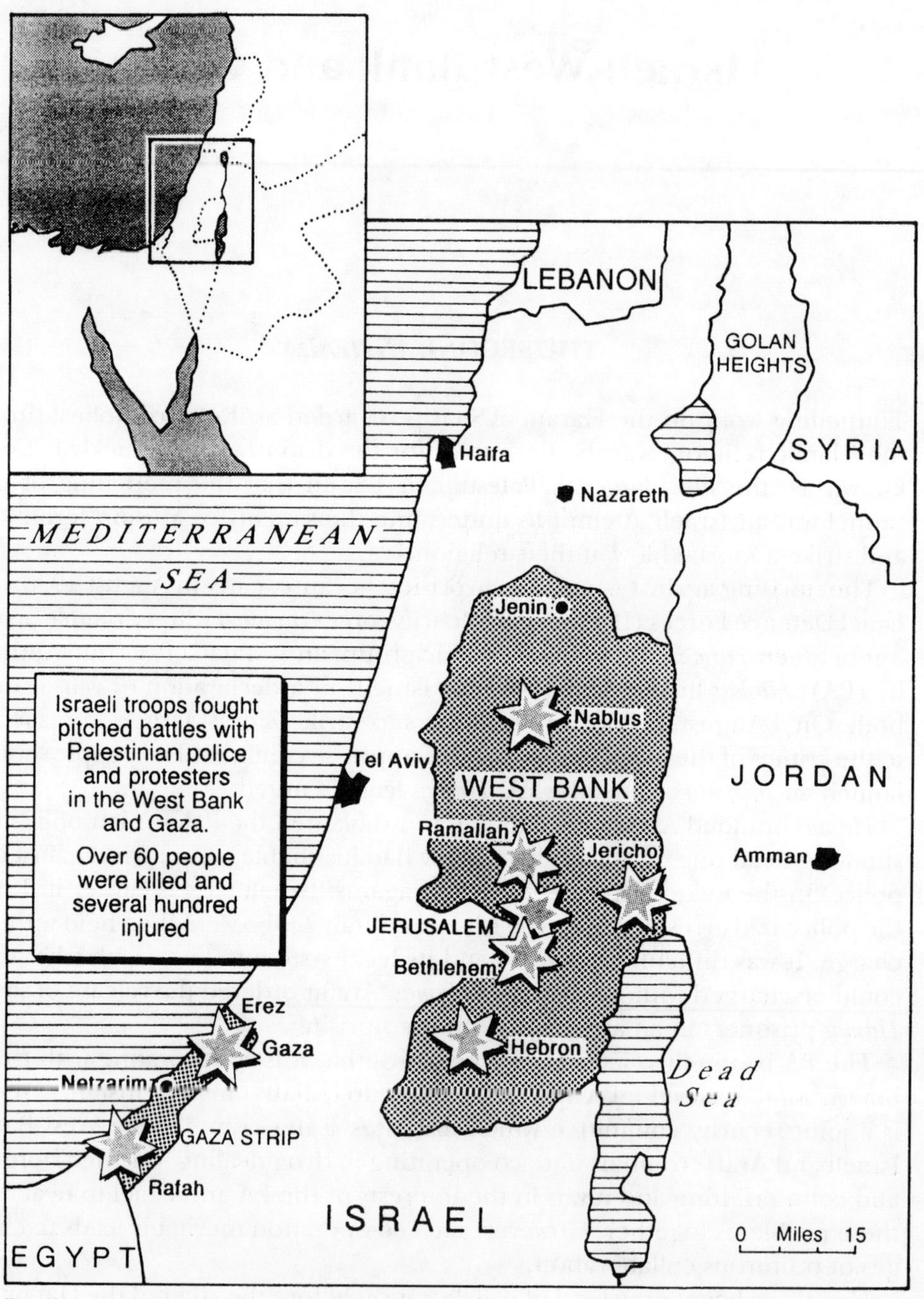

The Second Intifada: September 1996

area and cause of disputes. In 1910 a British nobleman bribed his way onto the Haram where he was caught hacking at the sacred rock with a pick axe. The consequent Arab riots went on for days.

After the Six-Day War of June 1967 the Israeli Ministry of Religious Affairs began excavations north of the Western (or Wailing) Wall, as well as building a synagogue and exploring the ancients walls under the Haram. In 1968 the ministry commenced excavation of the tunnel – known as the Hasmonean tunnel and that began the crisis nearly 30 years later.[2]

It should be explained that to the Western outsider of rational, pragmatic mind the affair generated passions out of all proportion to the problem. The so-called tunnel is actually an underpass 491 metres in length and it had been accessible to tourists for years. It had only one opening, and this, in the Jewish sector of the Old City, served as both entrance and exit. The municipal authorities had created an exit in the Arab Quarter in 1988, an act that caused so much unrest among Muslim Arabs that the authorities sealed it up.

In September 1996 it was decided to reopen this sealed off exit. The authorities said – but only after the *intifada* erupted – that they did this to allow more foot traffic through the narrow passage. Israeli security chiefs warned against the reopening but Prime Minister Netanyahu's advisors reportedly advised him that any danger had been exaggerated. Their mistaken way of avoiding trouble was to do the work at night and without consultation with the PA. The work was done on the night of 24 September.

When the Palestinian civil authorities heard about it they feared a plot was afoot to undermine – literally and metaphorically – their holy site. They also saw it as an Israeli attempt to prepare the ground to claim more of Jerusalem, whose final status was to be decided in the last stage of the Oslo Accords.

The Palestinians demanded the Arab eastern portion of the city as the capital of a Palestinian state and they were uneasy about the radically different hard line approach to 'peace' of Prime Minister Binyamin Netanyahu, compared with that of his predecessor, Shimon Peres. Netanyahu had delayed meeting Yasser Arafat and did so finally only under US pressure. He had refused to implement parts of the Oslo Accords already agreed by Shimon Peres, such as withdrawing troops from the West Bank town of Hebron and releasing Palestinian prisoners. Further, he had fanned Palestinian anger by announcing the building of 4,000 new homes in Israeli settlements on the West Bank.

If the Palestinians needed a pretext for an uprising, the Haram tunnel provided it. Some Western reporters permanently resident in Israel said it was as if Netanyahu had thrown a lighted match into a petrol tanker.

Initially, the Palestinian protests were spontaneous, with scores of youths on the Haram al-Sharif throwing rocks onto the Jewish worshippers below at the Western Wall. Elsewhere in East Jerusalem scuffles broke out between protectors and Israeli police. The protests grew into the new *intifada* first in Ramallah, but news of the tunnel 'outrage' spread quickly and in Gaza teenagers armed with stones and petrol bombs besieged Israeli checkpoints. Soon there was a front line, from where the young Palestinians maintained their attack. Others filled bottles with petrol taken from cars in the street and prepared them as 'Molotov cocktails'.

The next day, 26 September, Arafat and the PA approved the unrest. The Finance Minister, Muhammad Nashashini, made this clear when he told reporters that the ministers 'endorsed escalation of all means'.[3] They got more escalation than they could handle.

While the Israeli Command brought in heavy reinforcements Arafat and his lieutenants encouraged and incited unrest of all kinds. Several prominent men joined street marches, largely to ensure television coverage. Arafat's chief of Jerusalem Affairs, Faisal Husseini, was shown on TV and fainted as Israeli troops jostled him. Another of Arafat's key aides, Minister of Islamic Affairs Hassan Tahboub, was hit on the head, apparently with a club. Personally appearing on television, Arafat accused Israel of a 'big crime against our religion'. This comment poured more fuel on the fire.

But Arafat knew what he was doing. While he might have deplored the loss of life, Arafat the politician had much to gain from the *intifada.* Khalil Shikaki, the Director of the Centre for Palestine Research and Studies, based in Nablus, said to reporters, 'the Israelis threw Arafat a golden opportunity and he pounced on it'.[4] The opportunity Shikaki referred to enabled Arafat to:

- Improve his standing with his own people, who had lost much trust in him.
- Divert attention from the gross corruption and incompetence of the Palestine Authority.
- Distract public attention from his own increasingly dictatorial behaviour.

Most importantly, the *intifada* against the Israelis pre-empted the *intifada* that was building up against Arafat and the PA.

Soldiers of Arafat's PA made a show of restraining the youths but when the Israeli troops fired on the 'Molotov cocktail' throwers, first with rubber bullets and then with live ammunition, the uniformed Palestinians fired back. Yet only the day before they and Israeli soldiers had been making joint security patrols.

On Thursday 26 September the Israeli command brought up overwhelming force – attack helicopters, tanks and armoured cars and armoured personnel carriers. Casualties mounted and, as usual, many of the victims were people not involved in the fighting. The soldiers became victims of the inexorable law of civil commotion, that when firearms are available for use they will indeed be used.

On 'Bloody Thursday' – as the media were quick to call it – the conflict raged out of control. Palestinians, backed by Arafat's soldiers, fought Israelis – soldiers, police and civilians – anywhere on the outskirts of towns, in Jewish settlements in the West Bank and Gaza, at Jewish religious sites. In many cases the Palestinian attacks were suicidal. An observer, Ziad Abu-Amr, the political scientist and member of the Palestinian Legislative Council, said: 'People were just running towards death, attacking Israeli soldiers who they knew would shoot them.' The young Israeli troops were afraid of being hacked to death and some were killed in this way.

For the Palestinians, the worst Israeli offence occurred on Friday 27 September. Many Muslims were at prayer in the great Al-Aqsa mosque when youths threw stones at Israeli soldiers near the entrance. The soldiers opened fire. This was seen as an 'evil attack on the holy al Aqsa mosque' rather than as retaliation against stone-throwers. However it might be viewed, the incident was an incalculable public relations disaster. Several Israeli commentators were strident in their criticism of Netanyahu. In *Ma'Ariv*, Chemi Shalev wrote:

> During his traditional 100 days of grace since he took office, the prime minister has managed to overturn completely the new Middle East of his predecessor, Shimon Peres, and revive in its place the old Middle East. The international community is now united in its criticism of Israel, the Arab world fierce in its hatred and the Israelis and Palestinians are killing each other.[5]

An American journalist, writing in *Ramallah*, declared that a new description would be required for the confrontation raging between Israelis and Palestinians; 'There was no other word for it but war'.[6]

By the week's end Arafat had given orders to his troops to hold fire unless fired on but he was insisting that Netanyahu must produce meaningful concessions, such as closing the tunnel. In Gaza his language was much more inflammatory: 'Our blood is cheap in the face of the issue for which we are gathered here.' On Palestine radio an unidentified speaker said the time had come 'to slaughter all the Jews and to appoint a Caliph for Palestine'.[7]

Nevertheless, in the dying days of September the Palestinian police were trying to keep the still enraged mobs from engaging the Israeli troops. As an uneasy and unofficial truce took hold the casualties could be counted: 60 Palestinians and 14 Israelis killed and another 1,000 people of both sides injured. Some of these subsequently died.

In retrospect, it is possible, perhaps likely, that the violent confrontation was the inevitable result of allowing the PA to establish a military force far larger than that stipulated under the Oslo Accords. In September 1996 Arafat had at least 30,000 men under arms. He believed that he needed such military strength to protect himself and his administration against *Hamas* and dissident Palestinians who, while not Islamic extremists, accused him of betraying the Palestinian people by signing any form of peace agreement with Israel.

The new *intifada* had far-flung effects. In Cairo, the Arab league accused Israel of plotting the destruction of Islamic holy places. King Hussein of Jordan, Israel's only real friend in the region, said, 'Israel has violated the sanctity of the holy city'. Egypt warned that the destabilisation which followed the new *intifada* would provide fertile ground for extremist groups anxious to begin a new phase of international terrorism.

Appeals by men of goodwill have little or no effect in stopping the violence, as can be seen from an open letter by Israeli novelist Amos Oz to a Palestinian friend warning against the unravelling of the peace process. His letter was published in March 1996, more than six months before the new *intifada*.[8]

The Hebron Impasse

Whatever the terms finally agreed by Israeli, Palestinian and international negotiators for the Israeli withdrawal from 80 per cent of Hebron, as arranged under the Oslo Accords, the fact on the ground is that the withdrawal will bring an *increased* military presence in the city. It will be concentrated around the enclaves where about 50 Jewish settler families live amid 200,000 Palestinians.

These settlements have been transformed into armed fortresses; even the sentry posts are armoured in expectation of Arab attacks. The increased army presence is not only to protect the Jewish enclaves from Arab attacks but also to fend off violence by radical Jews. According to Shin Bet, the Israeli internal intelligence service, settlers could turn to attacks against Palestinians in order to wreck any eventual withdrawal. At the end of 1996 more than 1,000 Israeli soldiers were stationed in Hebron, more than two for every Jewish settler.

Ironically, there will officially be an Arab and an Israeli zone, thus creating a permanent divide between the two peoples. Under the Oslo self-rule arrangement the Israeli withdrawal was to have been completed by March 1996 but the Netanyahu government refused to implement this deal until it was satisfied that the Jewish settlers were secure.

The transfer of most of Hebron to Palestinian rule seemed to be imminent as 1996 ended but on 1 January 1997 an Israeli soldier shot and wounded eight Palestinians in an attempt to massacre civilians and sabotage the peace process. The soldier, who came from a Jewish settlement at Ma'ale Adumin, was arrested but the damage had been done.

The Palestinian chief negotiator, Saeb Erekat, blamed Israel over the incident, saying that it had put the security of the 400 Jewish settlers in Hebron above that of the scores of Palestinians. Other Palestinian officials, including Ahmed Tibi, an advisor to President Arafat, appealed for calm. But the Israeli security chiefs did not expect calm and ordered the settlers to stay in their homes because of fears that they would be attacked by Arab residents. Whatever the outcome of each incident, there will inevitably be many more. The cycle of violence is endless.

The depth of feeling over Hebron and its significance as a flash point for war can only be understood if its ancient history is explained. Religious Jews believe the cave of Machpelah, at the very heart of Hebron, was bought by the patriarch Abraham as a burial place for his wife. The Tomb of the Patriarchs is also a holy site for Muslims, who call it the Haram al-Ibrahimi; Ibrihimi is the Arabic word for Abraham. This is the fundamental core of the fanatical and bloody racial

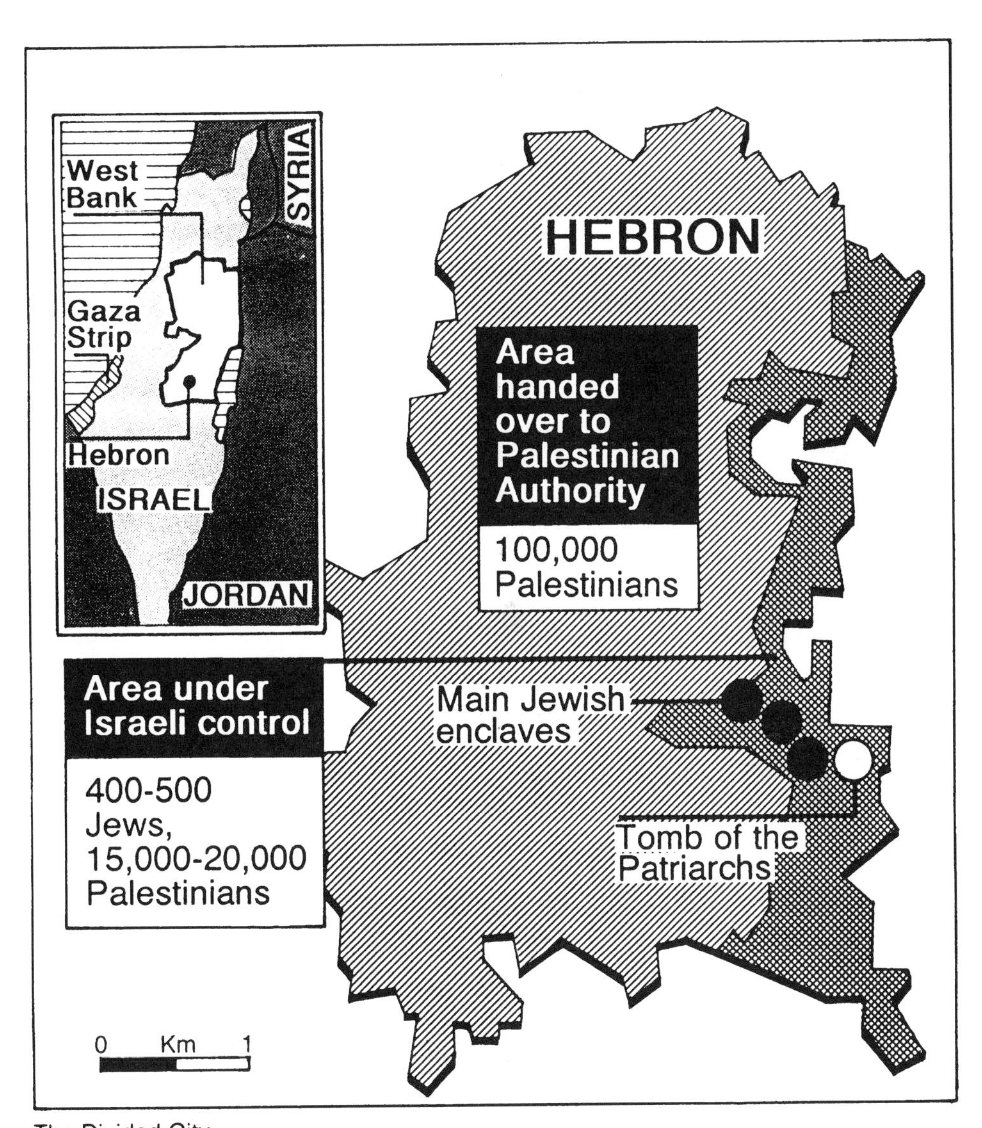

The Divided City

intolerance on both sides to the dispute.

In 1929, scores of Hebron's Jews were murdered after *riots* over prayer rights spread from Jerusalem. Each religion had, as it were, the right to pray at different times and at different places. These rules were often not observed. In February 1994, shortly after the Oslo Accords were agreed, Baruch Goldstein, from the nearby Jewish settlement of Kiryat Arba, massacred 29 Palestinians as they knelt in prayer in the al-Ibrahimi mosque.

I have never been in Hebron without feeling the seething discontent among the Arabs and the profound fear among the Jews, coupled with their determination to stay put amid a sea of enemies.

Hebron is the last of seven West Bank towns, occupied during the 1967 war, to be evacuated by the Israelis but it poses much more complex questions than the other six did. The Netanyahu administration believes that if it gives way over Hebron there will be demands for it to give way over East Jerusalem.

The IDF's Morale Declines

In one indirect but very real way the IDF suffered great damage as a result of the change of government from Labour (Peres) to Likud (Netanyahu), and its casualties in various ambushes and fights: its morale was seriously damaged. In a letter to Israeli newspapers on 13 October 1996 reservist officers and soldiers accused Netanyahu and his government of 'making every possible effort to drag us into an unjust war and destroy the hope of peace'.

Thirty-three reservists, ranging in rank from sergeant to captain and all from infantry and paratroop units, signed the letter. Technically, perhaps, the letter was mutinous but a former Chief of Staff, Rafael Eitan, now the Minister of Agriculture, said: 'We cannot ignore this letter. We need to think hard about the reasons the soldiers wrote it and then take care of them.'

The soldiers were forthright. 'Our fighting spirit has been torn from our hearts and without it we do not see how we will be able to participate and function in the next war. May it never come.'

An officer, speaking on Radio Israel, said, 'If the Israeli public thinks its army is strong and can strike whenever it wants, wherever it wants, they are wrong. It won't happen.'

Independent surveys at the end of 1996 showed that the government's failure to secure a comprehensive Middle East agreement was largely responsible for the erosion of military morale. Many Israelis were less inclined to do military duty and there was increased draft-dodging.[9] The army's remarkable position as a central unifying force was-under scrutiny. Reservists are the backbone of the IDF. Most men are required to spend one month a year on reserve duty until the age of 49.

Terrorist War Renewed

On 30 July 1997 two Hamas suicide bombers struck at Israel by exploding two 10kg bombs in a crowded Jerusalem market; 15 people were killed instantly and some of the 175 wounded died later. It was the worst outrage in a year. Hamas announced that the bombing was to avenge a crude poster drawn by an Israeli settler extremist depicting the prophet Muhammad as a pig. Other attacks would follow unless Israel freed all Palestinian prisoners within two days.

Yasser Arafat publicly condemned the bombing but this did not mollify the angry Israelis. Many suspect Palestinians were arrested, the peace talks were called off, access to and from Gaza and the West Bank was closed and payment of several million dollars owed to the Palestine Authority was suspended.

Arafat took no action against Hamas though there was little that he could do as he, too, is a Hamas target. The Palestinian police general Ghazi al-Jabali told journalists that if the Israelis went ahead with threats to conduct military operations against militants in the self-rule areas the police would 'fight to the death'. Jabali said: 'If they come here to Gaza or anywhere else they won't come out again'.

The scene was set for further confrontation – which is the aim of Hamas and its backers.

References

1. An Arabic word, *intifada* means 'a breaking free'. The first *intifada* began in 1987 and had lost much of its momentum by 1995. It is described in detail in *War Annuals 3–6*.
2. False allegations linking Israeli tunnelling to damage of the Muslim structures began in 1974 when Arab nations induced UNESCO to pass a resolution condemning Israeli excavations on the same tunnel. UNESCO appointed an expert group, headed by Professor Raymond Lemaire, its representative in Jerusalem, to investigate the accusations. Lemaire had been expected to find Israel guilty but instead he reported that 'the criticisms that have been levelled at the methods used in the excavations are groundless. The excavations near the Temple Mount are being carried out by a perfectly well qualified team of experts.'
3. Nashashini was quoted by Lisa Beyer, *Time Magazine*, 7 October 1996.
4. Beyer: *op. cit.* The analysis of the 'golden opportunity' is mine, not Shikaki's.
5. Quoted from *Ma'Ariv*, 29 September 1996.
6. Beyer; *op. cit.*
7. Arafat's comments and those of the speaker on *Palestine Radio* were quoted in the Israeli newspaper, *Ha'aretz*, 29 September 1996.
8. This is a short extract from the Oz letter: 'Israel in our homeland, Palestine is yours. Anyone who refuses to live with these two simple facts is either blind or evil. Two and a half years after signing the Oslo Peace Accords you and we still have to agree about terms for peace ... Yitzak Rabin was a brave man; he paid with his life for his effort [for peace]. Shimon Peres is a brave man: both his life and his political future are at stake. Now is the time for Arafat to demonstrate that he too is a brave man – or to give his place to someone who is braver than him (*sic*). ... The essence is clear and simple: we stop ruling over you and suppressing you and you recognise Israel and stop killing us. But up till now we have delivered and you haven't ... More Israelis are being killed by Palestinians after the agreement than before. ... Where are the Palestinian mass rallies against murder and in favour of peace?'

9. Details from *Jane's Defence Weekly*, 23 October 1996, and confirmed by reference to the Office of the Chief of Staff, Lieutenant General Ammon Shahak, who said that attempts to evade reserve duty had reached 'plague proportions'.

General As a general reference to the new *intifada*, see an article under that heading, by Miriam Shahin, in *The Middle East*, November 1996.

19

Israel and Lebanon

'GRAPES OF WRATH'

On 11 April 1996 Israel launched a campaign called 'Grapes of Wrath' against the Iranian-backed military terrorist movement *Hezbollah*, which is based in Lebanon. Despite progress in the 'Arab–Israel peace talks' and the obvious commitment of the Rabin-Peres government to peace, *Hezbollah* had been making numerous attacks against settlements in northern Israel. The day following the beginning of the Israeli offensive, the chief of military intelligence, Major General Moshe Yaalon, announced that 'Grapes of Wrath' would continue for 'as long as necessary'.

'The IDF [Israeli Defence Forces] have the persistence to continue an operation like this for a long time', he said. 'Any difficulty is more for the Lebanese than for us.'

First, AH-64 Apache gunships tried to hit *Hezbollah* leaders with 'precision' attacks but failed to locate their targets. Sites in south and east Lebanon as well as others in south Beirut then came under fire from F-16s, AH-1 Cobra gunships and the Apaches. For the first time since Operation 'Peace for Galilee' in 1982 the Lebanese capital was subjected to air attack. In all, the Israeli air force flew 1,200 sorties.

IDF artillery fired more than 10,000 shells against *Hezbollah* targets in southern Lebanon. About 60 people were killed and 200 injured, while 400,000 Lebanese fled their homes in the south, which the IDF declared a free fire zone. The IDF onslaught did not stop *Hezbollah's* rockets; more than 200 were fired into Israeli settlements. A rocket crew needed to be in action for only minutes before moving to another firing point and even the sophisticated Israeli tracking system could not pick up the *Hezbollah* men in time to counter-attack.

A second objective of Grapes of Wrath was to compel the Lebanese Prime Minister, Elias Hariri, to force *Hezbollah* to leave the southern region but without military help from the Syrian Army, which to a large extent occupies Lebanon, Hariri was helpless. Prime Minister Peres of Israel was exasperated and frustrated because in 1993 he had achieved a verbal understanding with

Hezbollah that it would end its rocket attacks against Israel and that Israel would not hit Lebanese civilian targets.

On 18 April 1996, Israeli shells hit the United Nations Interim Force in Lebanon (UNIFIL) base at Qana, southern Lebanon. About 600 refugees from the fighting further south were sheltering there and more than 100 were killed, with many maimed. Following international protests, Prime Minister Peres apologised and he, with other senior government and IDF officials, assured Israel's vociferous critics that the shelling was the result of a technical error. The deputy Chief of Defence Staff, Major General Matan Vilnai, told a press conference on 5 May:

> The main reasons for the mistake were an error of 200–250 metres in the exact location of the UN camp on our maps and a mistake in the system we used to measure the distance between the camp and the *Hezbollah* mortar site. From the centre of the UN camp to the perimeter was a distance of 70 metres which was not taken into account.

While the army's gunfire was the most destructive, the navy's Saar 5 and Saar 4.5 fast attack craft shelled Lebanese coastal targets while the air force flew bombing sorties against *Hezbollah* offices and depots. Phantom-2000s dropped four 2,000lb laser-guided bombs. TV coverage of the Israeli air attacks showed Apache and Cobra helicopters flying at about 5,000 feet but the chief of the air force, Major General Herzl Bodinger, said that wherever he could employ fighter aircraft he did so. 'The idea was to go after the infrastructure of *Hezbollah* so there was a lot of emphasis to destroy buildings and bunkers. Big bombs are needed for this job and they can only be carried by jets.'

Many of the air force attacks were filmed from the jets and helicopters. Journalists were shown film of *Hezbollah* Katyusha rocket crews in action, often close to a UN post. When they realised that they were about to be fired on from the air they abandoned their vehicle and launcher and ran for safety. The Katyusha has a range of more than eight miles but travels at 750 yards a second and has a mere 50º elevation from the ground. Even using the AN-TPQ 36 radar system the Israeli gunners could have no tracking accuracy.

The *Hezbollah* fighters often fire their weapons from beside a UN building. On 17 April a Fijian UN officer tried to prevent the *Hezbollah* from firing rockets close to his position on the coast road. A *Hezbollah* man shot him in the chest.

Operation Grapes of Wrath lasted 16 days and ended on 27 April. A total of 165 people were killed and 340 wounded, mostly Lebanese civilians. No Israelis were killed.[1]

A UN investigation was led by the senior military advisor of the Royal Netherlands Marine Corps, Major General Frank van Kappen, who came close to accusing the IDF of intentionally shelling the UN Qana base.[2]

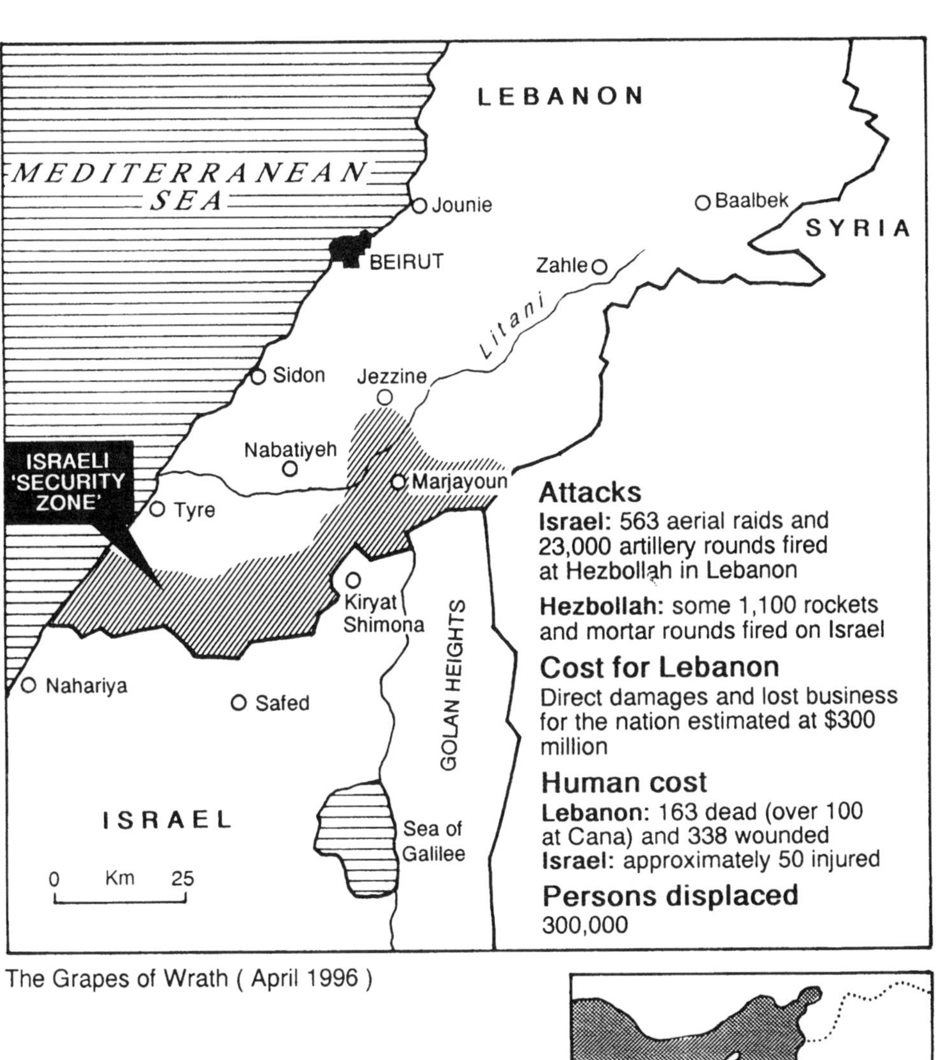

The Grapes of Wrath (April 1996)

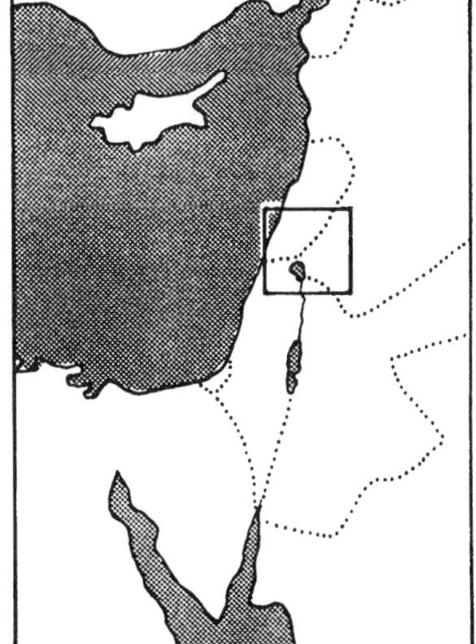

Ambushes and Punitive Shelling

Some of the most serious fighting between the *Hezbollah* and Israel took place in the middle of 1996. First several IDF men were wounded in an ambush in the security zone on 14 May. Within hours, Israeli artillery and warplanes repeatedly bombed *Hezbollah* strongholds in the Bekaa Valley and inside the eastern sector of the security zone. The attack and counter-attacks 'disappointed' the international diplomats who had negotiated a cease-fire agreement after the Grapes of Wrath campaign. But they could logically have expected nothing better to come from the cease-fire: under its terms the Shia terrorists were allowed to strike against Israeli troops and their allies the South Lebanon Army (SLA) inside the security zone, while Israel only had the right to self-defence. It was one of the more bizarre of the hundreds of cease-fire agreements reached in Lebanon over the years.

On 10 June, an Israeli patrol returning to its base near the ruins of the old crusader stronghold of Beaufort Castle was fired upon at 5.30 am. In the subsequent report issued to the press in Beirut, *Hezbollah* stated: 'A group of fighters ambushed a Zionist patrol consisting of more than 10 soldiers. The fighters hit them with machine-gun fire and rocket-propelled grenades and clashed with them until all fell to the ground, dead or wounded.'

The attack, a ferocious one even by *Hezbollah* standards, showed yet again the vulnerability of the IDF men in the security zone. Their attackers are rarely caught and it is known that many members of an ambushing party escape on motor scooters. Israeli artillery and aircraft went into action yet again, heavily attacking *Hezbollah* targets but there was no way of knowing if the Beaufort Castle attackers were hit.

On 6 August *Hezbollah* rockets killed one Israeli soldier and wounded others in their outpost in southern Lebanon. Three days later Israeli jets bombed a *Hezbollah* radio station which broadcasts from Baalbek and a nearby ammunition and fuel dump. The air raids took place two hours before the arrival of international observers to monitor the so-called cease-fire.

By now, *Hezbollah* attacks on the IDF were as intense as they were before the Grapes of Wrath retaliation. Significantly, the *Hezbollah* attack took place almost simultaneously with Syria's rejection of an Israeli proposal to resume peace talks, in what Israel called 'Lebanon First'. The Israeli air raid, while naturally a response to *Hezbollah* aggression, was also intended as a signal to Syria. The Syrian army of 40,000 controls the Bekaa Valley, the main area of direct military confrontation between Israel and Syria. The great majority of Lebanese would like the Syrians to leave so that the devastated country can be rebuilt and redeveloped. To this end the Lebanese would like a settlement with Israel. The Israeli raids into the Bekaa sent the message to President Assad, in Damascus: 'Control *Hezbollah* or we may attack its bases in Syria itself.'

In August 1996 the new prime minister of Israel, Binyamin Netanyahu, put forward a plan offering to withdraw Israeli troops from south Lebanon in return for security guarantees and the disarming of *Hezbollah* fighters. Lebanon's prime minister, Elias Hariri, peremptorily rejected the offer and restated Lebanon's demand for an unconditional withdrawal, in accordance with UN Security Council Resolution 425. Syria, too, rejected Israel's offer. Both Lebanon and Syria accused Israel of trying to drive a wedge between them. In the face of such obdurate mistrust it is difficult to see how progress is to be made.

The IDF's Special Operations in Lebanon

In February 1995 the IDF created a special unit, *Egoz* (Hebrew for walnut) to operate in a clandestine counter-terrorist role against *Hezbollah.* Its members, probably 100 of them, were selected from among volunteers of the Golani Brigade. By mid-year *Egoz* was operational, under the policy command of Major General Amiram Levine.

It was Levine who went public about *Egoz* late in 1996, an action so uncharacteristic of Israeli generals that it could be seen as a way of subjecting potential targets of *Egoz* to even greater terror. According to Levine, *Egoz* soldiers operate as military units, in uniform, but given its success rate and the difficulty of surviving in uniform in such a hostile environment his statement must be doubted. Levine also said that *Egoz* deserved public recognition as an elite unit.

He no doubt sincerely believed this but he was also intent on showing that the army was taking on *Hezbollah* in its own territory. This assurance was necessary because the army's morale was low. An Israeli outpost was overrun by *Hezbollah* fighters and the defenders fled. (See *War Annual 7*). In November 1996 an Israeli patrol was caught in a *Hezbollah* ambush, leaving five soldiers dead and eight wounded.[3]

The *Egoz* men were trained at Kiryat Tivon army base near Haifa. Other special forces have been trained there, including one that operates, often in Arab disguise, in the West Bank and Gaza. That *Egoz* has its kills is undoubted. In the second half of 1996 several senior *Hezbollah* men were killed in 'mysterious circumstances'. In all, 16 were killed during 1996. General Levine insists that *Egoz* does not operate against civilians, nor was it interested in arresting 'ordinary criminals'. Its business was exclusively against terrorists. During a press conference in December 1996, Levine said:

> *Egoz* is special because it concentrates exclusively on the fighting in Lebanon. Its professional level is high and it can operate anywhere, in the hill country, in thick undergrowth and in populated areas. Moreover, *Egoz* men are familiar with what kind of enemy *Hezbollah* is and the way it fights.

Some of the *Hezbollah* activities have been in response to *Egoz* operations. The commandos penetrated deep into enemy territory during the Grapes of Wrath operation in April 1996. In several places they supplied the information about the whereabouts of Katyusha rocket teams, so that artillery and air strikes could operate against them. According to one report, an *Egoz* patrol was unwittingly responsible for the Israeli shelling of the UNIFIL base at Qana. If this report is to be believed – and it is certainly credible – *Hezbollah* people spotted the Israeli patrol and attacked it. Trapped, the leader radioed for artillery cover. It was this fire that hit Qana.

Hezbollah – a profile

Hezbollah – meaning the Party of God – became active in 1982 when Lebanon was in turmoil because of a civil war and the Israeli campaign against the Palestinian terrorist organisations, Operation Peace for Galilee. Its leaders, Lebanese and Iranian extremists, wanted to create a fundamentalist Islamic state in Lebanon.

The organisation's finance, training and weapons were supplied by Iran and Syria and its members were trained in the Bekaa Valley by Iranian Revolutionary Guards of Lebanon. Its more senior ideological leaders were indoctrinated in Teheran and its intelligence chiefs were given instruction in Damascus by the Syrian secret service.[4]

Hezbollah's active members, those operating in the field are virtually all Shia Muslim; Sunni Muslims are not considered extreme enough in their views to make 'good' holy war warriors. In 1997 about 5,000 members were paid and armed, with much support from reserves. Within a week, *Hezbollah* could raise a force of 20,000.

Most *Hezbollah* leaders live in the southern suburbs of Beirut, others in Sidon. From here they direct the infiltration of villages and towns in the north of the 'UN-protected zone'. *Hezbollah's* active service area is Israel's security zone of Lebanon and the border areas themselves.

Hezbollah has never been short of weapons and ammunitions which reach them from across the Syrian border or from the sea. Israeli naval patrols capture or frighten off many arms smugglers but they cannot permanently seal Lebanon's southern coast. *Hezbollah* would be crippled if Syria were to withdraw its support but it pays the Syrian government to maintain a state of tension against Israel without being directly involved. Since *Hezbollah* fighters do not wear uniforms or insignia they are unidentifiable and merge so completely into Lebanese society that they appear to be ordinary civilians. Israeli intelligence knows their identity through spies and informers.

Hezbollah will never go out of existence of its own volition. While since 1982 it has become largely autonomous in its administration, it is the tool of Iran and Syria who will always find it useful in their war against Israel. Total Israeli withdrawal from Lebanese soil would not bring an end to *Hezbollah* activities, for *Hezbollah* would then become more aggressive against Israel itself.[5]

Deep Commando Strike

On 4 August 1997 commandos of the elite Golani Brigade made a raid north of their security zone in southern Lebanon. Israeli troops rarely thrust north into Lebanon and this raid was the deepest in eight years. It took the Golani commandos into thc hill rcgion ncar Nabatiych.

Helicopters landed the fighting men close to a *Hezbollah* base and a two hour firefight ensued. A brief army statement said that a number of terrorists were wounded, but casualties were heavier than this. According to a Lebanese source, the Golanis planted a bomb near the village of Kfour which killed five *Hezbollah* members and six civilians.

Significantly, the raid took place less than a week after the suicide bombing of the Jerusalem market, which was claimed by Hamas. Some Israeli Intelligence sources doubted the authenticity of this claim and suspected that *Hezbollah* infiltrators were the real perpetrators. The Golani operation could have been intended to show the *Hezbollah* leadership that its secrets were known to the Israeli military, who can strike with impunity deep into Lebanon. The raiders withdrew by helicopter before dawn, apparently without casualties.

References

1. The majority of the Lebanese living in the south of the country are Christian. They believe that in the dusty Hill town of Qana Jesus turned water into wine. They were saying after the massacre of 18 April that the holy wine had now turned into blood.
2. *The Times*, London, commented: 'The Israeli bombardment of the UN refugee camp near Tyre has highlighted the huge risks of using unguided heavy firepower to hit small targets close to civilian communities. The fear of "collateral damage" was one of the guiding principles which helped to focus the American-led coalition when it launched its massive air raids on Iraqi targets during the 1991 Gulf War. *Hezbollah* units launching Katyusha rockets and mortar rounds into northern Israel do not face the same restrictions. Their shells are fired indiscriminately and there is no concern about accuracy.'
3. In the midst of the truly bitter fighting between Israel and *Hezbollah* a remarkable exchange occurred between them. Privates Yosef Fink and Rahamim Alseikh had been reported missing in action against *Hezbollah* in 1986. In 1991 the families received word that their sons had perished in return for 51 *Hezbollah* prisoners and the bodies of nine *Hezbollah* men. There began the process of trying to have their remains returned to Israel for burial. As a nation, Israel is profoundly concerned about the return and proper burial of its soldiers, even of body parts. In August 1996 through a German-brokered deal, the bodies of Fink and Alsheikh were returned in exchange for the remains of more than 100 *Hezbollah* fighters and the release of dozens of prisoners.
4. In 1994, a total of 187 attacks against Israeli troops and positions by *Hezbollah* were recorded. There were 119 instances of artillery fire, 31 detonations of explosive devices and two frontal assaults on IDF positions. In 1995, *Hezbollah* committed a total of 344 attacks against Israeli troops and positions. There were 270 instances of artillery fire, 31 detonations of explosive charges and two frontal assaults on IDF positions. The year 1996 began in a similar fashion;

   28 February: Attempt by *Hezbollah* terrorists to infiltrate northern Israel utilising ultra-light aircraft.

   4 March: Detonation of explosive charge near Kibbutz Manara, in northern Israel, four Israeli soldiers killed, nine wounded.

10 March: Detonation of explosive charge in Israeli security zone in southern Lebanon, killing one Israeli soldier.

14 March: Ambush of IDF convoy on the Reihan–Aeyesha road. Eight soldiers wounded.

20 March: Suicide bomber detonates explosive charge in front of an IDF convoy.

30 March: Two Katyusha salvos were fired at the Galilee. One Israeli civilian was wounded.

9 April: Two Katyusha salvos were fired at the Galilee and 36 Israeli civilians were wounded.

10 April: *Hezbollah* mortar fire killed one Israeli soldier and wounded two others.

5. Lebanon is tied to the position and policy of its Syrian overlord. For Syria, *Hezbollah's* attacks in southern Lebanon are a useful card to take into its negotiations with Israel. *Hezbollah* keeps Israel pinned down and vulnerable in southern Lebanon, while Syria enjoys the political fruits and yet maintains a position of plausible deniability regarding the guerrilla attacks. Giles Trendle, in an article entitled 'Lebanon's Heavy Price of Allegiance', *The Middle East,* June 1996.

20

Liberia: Total Anarchy

When President Samuel Doe was assassinated in 1989 Liberia began a slide into instability and ruin. Various factions created 'armies', that of Charles Taylor being the most notorious and bloodthirsty. Liberia's alarmed neighbours tried on 12 occasions between 1990 and 1994 to bring about a cease fire but Taylor, confident that he would sooner or later be victorious, refused to compromise with Alhaji Kromah, leader of the Madingo ethnic group.

However, in August 1995 an apparent breakthrough occurred. With little preliminary fanfare, the US special envoy, Dane Smith, induced Taylor and Kromah to sign a ceasefire in the Nigerian capital, Abuja. It was known as the Abuja Accord. This was followed two weeks later by the setting up of a transitional government, with Taylor as its leader and Kromah as vice-president. In August 1996 presidential elections would be held.

The euphoria was remarkable and the general confidence about peace overwhelming. In Monrovia, capital of Liberia, Taylor announced, 'The war is over here and now'. Zimbabwean politician Canaan Banana, observer for the Organisation of African Unity (OAU), was equally positive. 'The sounds of violence will die', he said, 'and the bells of peace will ring forever'.

It was no surprise to foreign diplomats in West Africa that the Abuja Accord, and the Council of State which had been set up under it, fell apart within a month and bloody violence reigned again. 'Commander' Alexander Tweh of Taylor's National Patriotic Front of Liberia (NPLF) said to a Reuter's correspondent: 'We want peace but if anybody attacks us we will give them "pieces"'. As he spoke, he fired a rocket-propelled grenade at a position held by Kromah's men. Taylor and Tweh accused Kromah of spreading Islamic fundamentalism in Liberia and it is true that Kromah is an Islamic fundamentalist.

Under pressure from various politicians, including President Jerry Rawlings of Ghana, truces came into being, only to collapse. In the process a new rebel leader was thrown up – Roosevelt Johnson. Taylor refused to negotiate with Johnson, who was being sought on murder charges, and told him to give himself up to the United States embassy or the United Nations.

'As a government we are not going to negotiate with these terrorists', said Taylor. At the time West African peacekeepers were deployed around the barracks where Johnson's fighters were holding hundreds of hostages, including 22 soldiers of the West African peacekeeping force and about 50 Lebanese citizens.

Sixty thousand hungry, homeless Liberians wandered the streets of war-torn Monrovia. The Catholic cathedral was looted and the Archbishop of Liberia, Michael Francis, was robbed. The Archbishop joined the 2,000 foreigners ferried by helicopter to the US warships offshore. This was part of the US–European Command's Operation 'Assured Response', as the evacuation was known. By now – April 1996 – UN officials were reporting that some West African peacekeepers were joining in the looting of Monrovia, US forces established a support base at Freetown, Sierra Leone, north-west of Liberia, and evacuated more than 800 civilians by helicopter in the first five days of the operation.

The Americans considered it a major undertaking. The aircraft deployed came mainly from the US Air Force's 352nd Special Operations Group based at Mildenhall, UK. The Mediterranean Amphibious Readiness Group despatched to Liberian waters the USS *Guam,* USS *Trenton,* USS *Portland,* a destroyer and an oil tanker. On board the ships were 1,500 men of the 22nd Marine Expeditionary Group.

The 16-nation Economic Community of West African States (ECOWAS) arranged further peace talks for May 1996 but the leaders who were supposed to arrive in Ghana did not turn up. Fighting again broke out in and around Monrovia, with the rival factions attacking one another with heavy machine-guns, rocket-propelled grenades and small arms. Fierce fighting took place on the Johnson Street Bridge that leads into central Monrovia.

Despite the anarchy, the US ambassador to Liberia, William Milam, said that he still hoped that elections could take place by the end of 1996. 'I think the end of the year is possible if the hostilities were to stop and we were to get back to the peace process now', Milam said. 'But if we delay much longer, the end of the year becomes very problematic.' That the embattled ambassador, in the face of such intractable enmity between the factions, could make this statement was remarkable testimony to the optimistic nature of American diplomacy.

The fiercest fighting of April–May 1996 centred on the Barclay Training Centre, a walled military compound by the beach in Monrovia. It was taken over by 10,000 soldiers and followers of the renegade warlord Roosevelt Johnson, who held hundreds of hostages. The Barclay Centre might have been ideal for training but it was never meant to be a redoubt and it came under siege and fire from Charles Taylor's army. Each day dozens of people were killed by shellfire or died from the diseases which were rampant in the unhygienic surroundings.

By September, the UN was in despair over Liberia. Secretary General Boutros Boutros-Ghali warned the world in general, and the fighting factions in particular that the UN and African peacekeepers would pull out if there was no return to negotiations in a 'civilised atmosphere'. Reporting direct to the

Security Council, Boutros-Ghali said that Liberia was going the way of Somalia – into continuous degenerative ruin.

The fighting had even spilled over into neighbouring Côte d'Ivoire, (Ivory Coast) he said. Only 10 UN peacekeepers were deployed in Monrovia by the end of August, whereas there had been 100 in April. The others had fled for their lives. The relief agencies could not operate efficiently because of large-scale looting of organisational property and supplies. By September UN troops and agencies had lost a total of 489 vehicles, worth well over $8 million.

Despite the chaos and savagery, the murders and poverty, Liberia did have one positive development in September – it acquired its first woman head of state, Ruth Sando Perry, aged 56. This was one of the results of Liberia's 14th peace agreement, concluded in Abuja. One of its provisions was that 60,000 fighters would be disarmed before the end of January 1997, presidential and parliamentary elections would be held by May 1997 and a new government installed in June.

Charles Taylor made life difficult for Mrs Perry because he could not buy her with gifts. She attempted to use government assets to pay civil servants' salaries, which were 10 months in arrears, but Taylor blocked this move.[1]

By UN estimates, at the end of February 1997, 200,000 people had lost their lives in the Liberian fighting and more than half of the nation's 2.8 million had fled their homes. The economic damage was so great that it could not even be estimated.

However, that same month Mrs Perry said, 'I have the personal feeling that the war is over.' Her confidence followed a disarmament operation which began on 22 November 1996 and ended on 31 January 1997. More than 23,000 combatants from all factions handed in their firearms. According to Major General Victor Malu, commander of the West African peacekeeping force, this figure represented more than 70 per cent of all the factional fighters at large across the country. Foreign diplomats say that the true figure is about 40 per cent. Nevertheless, Malu publicly announced that he would use 'all means available' to search for and seize any remaining weapons. After a certain time anybody caught with a weapon would be treated as a criminal. With the ECOWAS countries apparently determined to end the violence in Liberia, after several years of war the prospects for peace seemed brighter than ever before.

Liberia's Boy Killers

In a special report to the UN Security Council, in November 1996, Boutros Boutros-Ghali, stated that about 20,000 child soldiers were under the control of the six major warring factions of Liberia. UNICEF (the UN children's fund) was trying to get children off the battlefield, he said.

UNICEF has not been noticeably successful, though another agency, the Children's Assistance Programme in Monrovia, has brought some boys into a rehabilitation centre in the capital. The process is not easy because the warring adult leaders, who are merciless and ruthless, know that their child soldiers – many of whom are below the age of 10 – are totally loyal, militarily efficient and largely without fear. Indoctrination has eradicated any moral qualms they might have had. Ordered to fight to the last 'man' in some desperate encounter, they will do so. Many a battle has taken place between boy fighters on both sides. In a great push by Charles Taylor in 1992, boys high on drugs fought savagely in the swamps around Monrovia.

Armed with AK-47s, the boy soldiers feel invincible and they are reluctant to give up their weapons to the adults of the Christian missions and the Children's Assistance Programme. The offer of classes in trades, which will bring them a skill and a job, seems an inadequate exchange for the surrender of their beloved firearm. They feel that they are being put in a cage. Officials of the Children's Assistance Programme fear that many of the child soldiers, most of whom have committed atrocities, are the criminals of the future, even in a reformed Liberia.

A coup for 'the sake of stability' – Sierra Leone

Late in May 1997 Liberia's neighbour, Sierra Leone, another impoverished West African country, was the scene of a violent coup. The coup leader was Major Johnny Koroma, head of an Armed Forces Revolutionary Coucil. His stated aim was to bring about the reinstatement of the democratically elected president Ahmed Telan Kabhah.

However, the UN, the Organisation of African Unity and other bodies supported the government-in-place. A West African peace-keeping force was already in Sierra Leone and it was reinforced by Nigerians following Koroma's coup, which had left more than 100 dead. His ill-disciplined troops were looting and harassing civilians. Many of these men were former rebel fighters who poured in from the bush after Koroma's coup.

Koroma was hoping that the UN Security Council would do what no other international body had done – that is, rule out the use of military force to remove the soldiers from power. Koroma, a shrewd politician, reopened the country's borders with Liberia and Guinea and staged a rally which he said was for a peaceful resolution to the crisis. Nevertheless, Nigerian troops bombarded Koroma's positions. The situation remained tense throughout 1997.

References

1. A Western diplomat said: 'Since 1990 Liberia has had a succession of token leaders. The warlords have deliberately appointed weak and ineffectual civilians. They become puppets or, when things go wrong, they are scapegoats. The same fate will befall Mrs. Perry, even though she is genuinely neutral with a rational, humanitarian mind. In Liberia these are failings, not qualities to be admired.'

21

Libya: An Islamic Backlash

Following a failed military coup in 1993, Libya has been a cauldron of numerous revolts, prison riots and breakouts as well as assassinations of ministers and senior officials. Much of the unrest is covered up by the government-controlled media.

Unrest escalated into outright war as Muslim militants, preaching a religious revolution against the Gaddafi regime, established bases and hideouts in the eastern part of the country as opposition to President Gaddafi widened. The military offensive against the rebels began in July 1996, largely under cover of a large-scale and lengthy live-fire air and ground exercise. This took place in the Jebel Akhdar region near the coastal city of Derna, 200 miles west of the Egyptian border. (Jebel Akhdar – or Green Mountain – was the scene of fighting between the British and the German–Italian forces during the Second World War.)

The government has never officially acknowledged any rebel activity because Gaddafi likes to give the impression that his 'Brother Muammar' regime is popular. However, diplomatic sources say that during 1996 750 service personnel and armed militants were killed and another 1,000 wounded. Among the casualties was Muhammad al-Hamy, leader of the Martyrs Movement, one of the several Islamic groups active against Gaddafi.

Air force fighter-bombers and helicopter gunships repeatedly raided rebel targets but apparently with little effect. Just how many air force planes were involved is unknown; many aircraft are known to be grounded because UN sanctions have caused a chronic shortage of spare parts.

Anti-Missile Defence

While facing internal political difficulties, Gaddafi was simultaneously trying to develop anti-missile systems because of American threats to attack Libya's chemical weapons plant inside mountains of the Tarhunah Range.

Throughout 1995 and for much of 1996 the Clinton administration was remarkably explicit in it threats, through senior officials, intended to signal US determination to block Libya from completing the chemical weapons complex at Tarhunah. Clinton and his cabinet were pursuing the problem along two tracks – pressing diplomatic efforts to frustrate Gaddafi's plans while developing military alternatives in the event that diplomacy failed.

Harold Smith, who was at that time responsible for the Pentagon's nuclear, chemical and biological weapons' strategies, said that the US needed at least two years before the military had a non-nuclear weapon capable of destroying targets as deep as the Libyan plant. 'We could not take it out of commission using strictly conventional weapons', he said. 'But a deep-penetration non-nuclear warhead is in an advanced stage of development.'

By repeatedly going public with its accusations against Libya the Clinton administration was trying to increase pressure on European and Asian countries to halt shipments of specialised construction equipment and other materials that Libya needed to complete its underground complex.

The publicity given to Libya's anti-missile programme is one way for Gaddafi to divert his people's attention away from the country's economic ills. However, a report from the official Libyan news agency, *Jana*, that Gaddafi would send 'millions of Muslim' to protect Tarhunah should be taken seriously. Libya has a total population of only four million but Gaddafi would certainly send as many people as possible to the area as a human shield if he seriously expected an American attack. He knows that the prospect of civilian casualties on a large scale would deter the Americans.

The Clinton administration would certainly like to destroy the Tarhunah complex, which, according to the CIA, is the largest in the world. It covers six acres and cost £350 million. However, the decision has been made not to use nuclear weapons. Instead, deep-penetration conventional weapons are being developed to put the plant out of operation.

In August 1996 two German businessmen were arrested in Munchen-Gladbach on suspicion of smuggling equipment for the manufacture of poison gas to Libya. The equipment was the automation system, which mixes poison gases, including sarin and soman, two of the most lethal substances. Sarin was developed by the Nazis and was used by terrorists in the Tokyo underground railway system in 1995.

While denying that Libya is producing chemical weapons, Gaddafi repeatedly asserts the right of Arab states to possess them in order to counter Israel's regional monopoly of nuclear weapons.[1]

In August 1996 the US, which has banned trade with Libya since 1986 with the aim of forcing Gaddafi to hand over two of his agents indicated by the US and British authorities for bombing Pan Am Flight 103 over Lockerbie, Scotland, in December 1988, further increased its efforts to damage Libya's economy. President Clinton signed a bill under which non-US firms which invest more than $40 million a year in Libya's oil and gas industry can be penalised. This figure might seem like big spending but it is much less than would be required for major investment in Libya.

The US actions have had the effect of intensifying the revolt by a new generation of Islamic zealots. These people are even more aggressive in challenging the regime than Gaddafi and his followers were when they challenged the monarchy in 1969.

References

1. Gaddafi claims that the tunnels are part of his 'Great Man-Made River Project', to carry water from southern aquifers to coastal cities and to take another flow of water from the Mediterranean to create a large lake which might change the climate of part of the Sahara Desert. However, US satellite photographs convince engineers that they are looking at a chemical warfare plant.

22

The Middle East: Revolution of Rising Expectations

The Middle East is both a political and geographical term and one overladen with connotations of discord in the religious, sociological and economic senses. For the British, the Middle East covers all the north African Arab countries from Morocco to Egypt, as well as the eastern end of the Mediterranean, Israel, Lebanon and Syria. It even extends further to include Jordan and Saudi Arabia. Many British geographers and political commentators include Turkey in the 'Middle East' and most would certainly name Cyprus. For Americans, Morocco and Algeria are part of the Near East. The French have another division: while Morocco and Algeria are part of their *Moyen Orient*, Israel, Lebanon and Syria are the Levant.

Since 1948, the year of the establishment of the state of Israel, the media has referred *ad nauseum* to 'the Middle East dispute'. Most journalists mean by this term the succession of wars fought between the Arab states and Israel. But disputes are endemic in the Middle East and have been for centuries. Civil wars within Arab countries and between Arab nations run into hundreds, to say nothing of the wars waged by the great powers in the Middle East.

At one time many politicians, military planners and writers would have included Persia (now Iran) and Mesopotamia (now Iraq) in the Middle East but modern Iran and Iraq are now firmly classified as 'Gulf states', together with all the much smaller Persian (or Arabian) Gulf countries.

However 'the Middle East' is defined, it is a cockpit of conflict in which nations tear themselves and one another to shreds – or threaten to do so. It is a region without trust or compromise and its tensions and disputes draw in the remaining superpower, America, and the traditional powers, Britain, France and Russia, as well as United Nations peacekeeping forces. In the Middle East no dispute is isolated but is interconnected with others.

The better to clarify and describe the actual wars, the close-call wars and the incipient wars, I define the Middle East as consisting of Libya, Egypt, Israel, Jordan, Saudi Arabia, Lebanon and Syria. I cover also the military activities of Iran in Lebanon but Iran of the ayatollahs is more directly involved with the destabilisation of the Gulf.

Iraq's belligerence impinges upon Syria, Jordan and Israel but its war against its own Kurds and plans for a renewed war against Kuwait are of even greater concern and are dealt with separately.

The language of violence is normal and natural in the Middle East and terrorism is increasingly a form of war. For instance, various organisations, groups and 'movements' – now the more common term – attack and kill in order to prove that they are forces to be reckoned with. The many Islamic fundamentalist movements, lacking real political power, make up for it with violence. In April 1996 Muslim militants massacred 18 tourists at a Cairo hotel. The shooting followed warnings by *Hezbollah* in southern Lebanon that they would attack Israelis abroad as well as at home. In fact, most of the victims at the Cairo hotel were Greek, but the hotel was known to be popular among Israelis and the Egyptian police suspected that the three gunmen, armed with AK-47s, and their woman companion, armed with a revolver, believed they were killing Israelis.

The massacre was also a warning to the Egyptian government of President Husni Mubarrak, which in March had been host to a world summit on ways of dealing with terrorism. Egypt suffers more from terrorism than any Arab country other than Algeria.

Syria engages in proxy terrorism. It encourages *Hezbollah* and Palestinian groups to kill Israelis in order that Israel will understand that its very existence will be tense and dangerous unless it agrees to Syria's terms for peace.

A British journalist, Martin Woollacott, describes what he calls 'the stereotypical Israeli situation' in relation to violence. He wrote, 'You kill people in order to send a message to another government that it should use violence against the people who are using violence against you, without real expectation that it will work, but in order to prove to your people that you are doing what you can.'[1]

Many leaders in the Middle East say openly that they are forced to make war in order to make peace. Any Israeli leader who does not react forcefully to aggressive acts by *Hezbollah* – such as firing Katyusha rockets into northern Israel – would soon lose the support of the electorate.

Hezbollah and *Hamas*, which have much in common, regard themselves as at war and, indeed, long ago they declared *jihad.*

The Middle East is a violent place partly because the Arabic language is full of rhetorical violence. When Saddam Hussein promised 'the mother of all battles' during the 1990 war with Kuwait he was perfectly understood by the entire Arab world, even if the phrase was theatrically extravagant to Western ears.

Within the Arab world great changes are coming from below because of the revolution of rising expectations and because of powerful demographic forces. In 1967 Egypt had a population of 30 million; in 1997 it had risen to 60 million and is growing at the rate of more than one million a year. The urbanisation of this growing population, not only in Egypt but throughout the Arab world, creates explosive tensions, as the bloody example of Algeria shows. (*q.v.*)

Since Iran's revolution its population, too, has doubled and while the ayatollahs control the military power much has changed in the country. The mosques, which were once full, are now empty – proving the decline of the awe and fear in which the ayatollahs and mullahs were held. Western culture has revived among the young and anti-Americanism is dying except inside the religious seminaries and important government ministries.

Future leaders in Teheran may have a different attitude to Israel from those now in power. It would be the attitude of the late deposed Shah and his forebears, whose political strategy was to maintain an alliance with the Jews of Palestine in order to find support against the overwhelming number of Arabs, whom the Iranians hate. The earlier Gulf War between Iraq and Iran was often depicted as a strange falling out of Islamic brothers. But Islam was never a binding force between Arabs and Persians (Iranians).

Corruption is endemic throughout the Arab world and the people know it. Despite enormous social changes nothing has happened in terms of political reform or evolution. The masses have no respect for regimes, many of which are dominated by an autocratic or religious minority. In Syria, for example, the Alawites number only 10 per cent of the population but controls the government and the armed forces.

Only one Western-style democracy exists in the Middle East and that is Israel. Although the Arab states have parliaments, it is the presidents, princes and monarchs who have overriding authority. Kuwait is a classic case: while a parliament of a kind exists, the country is run by the Al-Sabah family.

Against the background of corruption and incompetence, burgeoning population growth, poverty, Westernisation and rising expectation, violence is ever more likely and in all the countries of the region. Since the publication of *War Annual 7*, there have been assassinations and attempted assassinations, riots and bloody demonstrations, terrorist violence and State counter violence. This background spawns wars.

In the Middle East, war can so easily be talked-up. The Syrian President Hafez al-Assad indulged in extravagant war talk in August 1997, vowing to take the Golan heights by force if negotiations with Israel failed. 'Liberating the land is a sacred duty which we will not hesitate to perform and to sacrifice ourselves for,' he said on the 52nd anniversary of the establishment of the Syrian armed forces.

Such language – and al-Assad's speech which went on for three hours – inflames the expectations of the armed forces, especially of the generals. At the ceremony, many generals urged the president to declare war *now*.

References

1. *The Guardian*, London, 15 April 1996, and widely syndicated. The article was aptly titled 'Trail of Blood on the Road to Peace'. Woollacott said, 'Whatever their mother tongue, the people of the Middle East are all fluent in the region's second language, that of violence'.

23

Northern Ireland Terrorist War

BACK TO BASICS, BACK TO BOMBS

In *War Annual 7*, I altered the classification of the conflict in Northern Ireland from that of open war to suspended war but pointed out that the war was still going on in the minds of men and women addicted to violence. Also, I stated that the Irish Republican Army (IRA) and Sinn Fein were indivisible. In several parts of the world terrorist organisations maintain a 'respectable' political front through which they can make a show of distancing themselves from the active terrorist operations. Yasser Arafat did this as leader of *Fatah*, an avowedly terrorist group, and the Palestine Liberation Organisation (PLO), of which he was the chairman. As *Fatah* chief he would not have been welcome in any peaceful, democratic forum but as chairman of the PLO he succeeded in achieving recognition as a statesman. He was even received by the Pope.

Gerry Adams, leader of Sinn Fein, has not been quite as successful as Arafat but he has shaken hands with the President of the United States. He could hardly have expected such a propaganda triumph as a leader of the IRA.

The Chief Constable of the Royal Irish Constabulary (RUC), Sir Hugh Annesley, has stated that the IRA and Sinn Fein are 'inextricably linked' at the leadership level and Adams was the 'driving and controlling force' in both organisations.[1]

When he was 20 years old, Adams was one of the first people to volunteer to join the Provisional IRA after the organisation was set up in 1969. Between 1973 and 1976 he was interned, a period that served to increase his status in Sinn Fein/IRA. Towards the end of the 1970s, Adams was heavily involved in producing a 'prospectus' for the IRA.

It was Adams who produced a document called *The People's Assembly* which became the basis for 'trials' and punishments of its own members as well as people within the Roman Catholic community suspected of informing for the British. In 1994 he produced a document entitled *Tactical Use of Armed Struggle*, which was read out at IRA meetings in April that year. This important document disclosed that the 'peace process' to which Sinn Fein gave public support was only a tactical expedient. It offered a better method, at the time, of bringing about the IRA's aim of a 32-county republic for the whole of Ireland. A return to violence was neither then nor later ruled out.

Liam Clarke, one of the best known authorities on the workings of the Sinn Fein IRA, has revealed details of Adams's staff and senior associates. He says that Adams trusts few people whose loyalty has not been proved in the IRA.[2] Among his close associates:

- Siobhan O'Hanlon, his secretary, is a convicted terrorist who took part in the planned attack on Gibraltar in 1988 which led to the killing of three of her associates.
- Richard McAuley, his personal press officer, is a former bomber.
- Gerry Kelly, his close confidant, escaped from the Maze prison.
- Joe Cahill, Sinn Fein's head of finance, is a convicted murderer and gun-runner.
- Paul 'Chico' Hamilton and Jim McCarthy, Adams's bodyguards, were kidnappers. According to Clarke, they kidnapped Martin McGartland and took him to a third-floor office for an IRA 'courtmartial', accusing him of having been a police informer. McGartland escaped by jumping out of the window; he survived the fall.

VIOLENT END TO THE CEASEFIRE

The peace process began in 1993, through approaches made by the moderate republican politician and leader of the SDLP, John Hume. The British Prime Minister, John Major and his Irish counterpart, Albert Reynolds, became involved and after tortuous negotiations the Downing Street Declaration was produced. Hailed as the best chance for peace in 25 years, the agreement had at its core the idea of bringing Sinn Fein into dialogue with Northern Ireland's other parties. This was supposed to end the IRA's isolation and its justification for armed struggle. Also, it was hoped that Unionist (or Loyalist) paramilitary groups would join the ceasefire. Eight months after the Downing Street Declaration a cease fire was agreed to.

All the parties concerned, the British government, Unionists and Republicans, took up fixed positions. The IRA was called upon to hand over its weapons for 'decommissioning', a precondition that was never remotely likely to be accepted. For the IRA decommissioning weapons was nothing less than military surrender.

With the collapse of the agreement imminent, US Senator George Mitchell, who had been brought in to chair the talks, suggested that decommissioning should be shelved and somehow sorted out during negotiations. The IRA and the Dublin government opposed the plan but the Unionists supported it. Prime Minister John Major seized on another of Mitchell's ideas – an election in Northern Ireland. The Irish government and the IRA opposed this but the Unionists supported it and Major insisted on it, even when MI6 warned him that the cease fire was about to end.

It ended on 9 February 1996 when the IRA detonated 450kg of explosives in London's docklands. The blast killed two people, devastated 90,000 square yards of office space and caused £200 million worth of damage to the showcase development of South Quay's Canary Wharf. The ceasefire had lasted 17 months. The bombing of Canary Wharf could not have been decided on and mounted in a matter of only a few days. According to a British intelligence report, Sinn Fein/IRA must have known about the attack for several weeks, even as they were negotiating peace.

A few miles east of Tower Bridge and the City of London, the Canary Wharf tower block is the tallest building in Britain. It houses major finance and communication industries, media organisations, banks and brokerage firms. Overall, the annual business done at Canary Wharf was estimated, at the time of the bombing, to generate £25 billion annually.

The Canary Wharf complex was always likely to be an IRA priority target because a bombing there would produce international media coverage – highly desirable to the IRA and Sinn Fein. The area has only low density housing so there would be no heavy casualty toll among 'ordinary people'. By striking at the banking, commercial and business establishment, the terrorists hoped to be credited with humanitarian principles.

During the ceasefire Britain and Northern Ireland had become increasingly relaxed. The Canary Wharf bombing ended all that, in both a civilian and a military sense. David O'Reilly, writing in *The Canberra Times*, described the effect on the British people in prose as graphic as any of the world's journalistic reactions:

> In recent weeks the terrorist bombs planted in London have rekindled fear, frustration and uncertainty in an already battered nation. Indiscriminate murders without remorse, a populace held hostage by the hidden threat of sudden injury, blindness or disfigurement from exploding shrapnel (*sic*) and slivers of glass; one of the world's great cities ambushed and then brought to a halt, its economy and reputation buffeted, the lives of millions of its citizens disrupted and stressed.[3]

However, it was not only the British who were suddenly again at war. Within hours all the overt evidence of conflict was back in place in Northern Ireland. Well over 500 more troops were flown into Belfast for deployment in South Armagh. Soldiers and police manning roadblocks at the ports and airports were again wearing flak jackets, and military vehicles, rarely seen for nearly 18 months, were suddenly everywhere in evidence.

The effects of the London bombing were also felt in the Republic of Ireland. The Irish army and the police had stepped down security operations on the border between the Republic and Northern Ireland, transferring two border-security helicopters and two bomb-disposal units. In the wake of the bombing all were hurried back to the border. The regular army was to have been reduced from 13,000 to 11,500 but the reduction was deferred. The Irish navy's fleet of seven patrol vessels was strengthened to make it more difficult for the Sinn Fein/IRA to smuggle weapons into the country.

A few days after the Canary Wharf attacks a 4lb bomb was found in a telephone box in Charing Cross Road, London, and a large part of the city was shut down so that it could be defused. This was followed by the death of an IRA terrorist, Edward O'Brien, when a bomb he was carrying on a London bus exploded, injuring scores of people. Police found an explosives factory at the dead terrorist's home in South London.[4]

The Annual Orgy of Violence

Trouble between Loyalists and Republicans flared in July when the 'marching season' was in full swing.[5] In Portadown, Loyalist demonstrators provoked angry recriminations from the town's Roman Catholic minority. On 7 July 1996, in an effort to prevent riots, the police ordered their march to be rerouted. The main riots came not from the Catholics but from the Loyalist Protestants, who demonstrated all over Northern Ireland.

Two extra battalions of British soldiers were brought in, bringing the army's total presence to 18,000 men, the highest number since 1982. Even then the army was stretched thin to cope with the many crises.

Catholics then went wild, especially in Londonderry where about 1,000 bombs (including petrol bombs) were thrown at the police and soldiers. The security services responded with volleys of plastic bullets as they had earlier done against the Protestants. In Belfast, a police station came under gunfire, the second incident of armed sectarian attack on the police since the IRA ceasefire.

The worst terrorist outrage took place on 14 July at Enniskillen, 100 miles south-west of Belfast, where a bomb destroyed a hotel and injured two people. Elsewhere three policemen were shot by snipers.

The Intelligence Campaign Against the IRA

Dame Stella Rimington, Director-General of MI5, believed that to campaign effectively against the IRA better co-operation was needed between the police and MI5 in counter-terrorist operations. As a result, MI5 took over as the lead agency on the British mainland for intelligence-gathering against the IRA. When Dame Stella retired in 1995 her successor, Stephen Lander, continued her work in convincing doubting police officers that the policy was effective.

A good relationship developed between Commander John Grieve, head of Scotland Yard's Anti-Terrorist Branch, and Assistant Commissioner David Veness, in charge of specialist operations with the Metropolitan Police, together with senior MI5 officers. Improvements in the quality of intelligence were quickly apparent. In September 1996 a major operation was mounted and on 23 September a number of IRA suspects were arrested and 10 tonnes of explosives found. The operation, in London, Surrey and Yorkshire, also found some booby devices, an indication that attacks were to be made on people whose names were found on a captured 'hit' list.

In July 1996 a surveillance team from MI5 and Scotland Yard tailed Diarmid O'Neill, who was considered to be a dangerous and important IRA suspect, to the entrance of the Channel Tunnel. It is believed that the IRA had considered blowing up the tunnel but that this was abandoned because of logistical difficulties. O'Neill and a companion were then seen near Folkestone, examining the layout of the tunnel's electricity system. Unaware of the surveillance O'Neill scouted other possible targets. The security forces, spearheaded by marksmen from Scotland Yard's SO19, planned a raid on O'Neill's home in Hammersmith, London, on 23 November. During the final stages of the briefing these specialists learnt of O'Neill's suspected involvement in an IRA plot to plant huge truck bombs throughout London. He had visited a security warehouse in North London, where the IRA had stored 10 tonnes of fertiliser used for making explosives. During the storming of his flat O'Neill, aged 27, was shot dead. Surprisingly, no weapons were found in his flat. Because O'Neill was unarmed and apparently did not actively resist arrest, his death gave the IRA a propaganda advantage.[6]

The IRA: Weapons and Structure

The Provisional IRA (or Provos) is a small group of about 400 active volunteers, but they are highly trained, well-organised and well-equipped. British intelligence and security organisations have reported that the IRA has northern and southern commands controlled by a central 'army council' – which has two Sinn Fein members – and that it is ultimately responsible to an overriding 400-member General Army Convention (GAC). Some of the several thousand IRA supporters and sympathisers belong to this Convention.

Generally four to six active service units operate in Britain, some of them on a long-term basis. They seem to have limitless finance and they have caches of Semtex explosives and weapons on the mainland. An MI5 report early in 1996 estimated that the IRA possessed more than 1,500 rifles and machine-guns and one million rounds of ammunition. However, the quantity is immaterial; the IRA chooses its targets carefully and uses only a small quantity of its war resources.

The IRA stock of weapons is so very large and varied that it makes sense only if Sinn Fein/IRA plans a major uprising or civil war. The only alternative explanation is that weapons have an hypnotic attraction for the IRA. Mere possession of weapons gives them a feeling of power. According to *Jane's Intelligence Review*, July–August 1996, the IRA had these weapons in its armories:

> 40 RPG-7 rocket launchers; 20 12.7mm heavy machine-guns; 650 AK47 assault rifles; 'a few dozen' Armalite assault rifles; one Barrett M82A1 sniper rifle; six flamethrowers; three tonnes of Semtex and 600 detonators. Other sources show that the IRA also has 10 Sam-7 surface-to-air missiles.

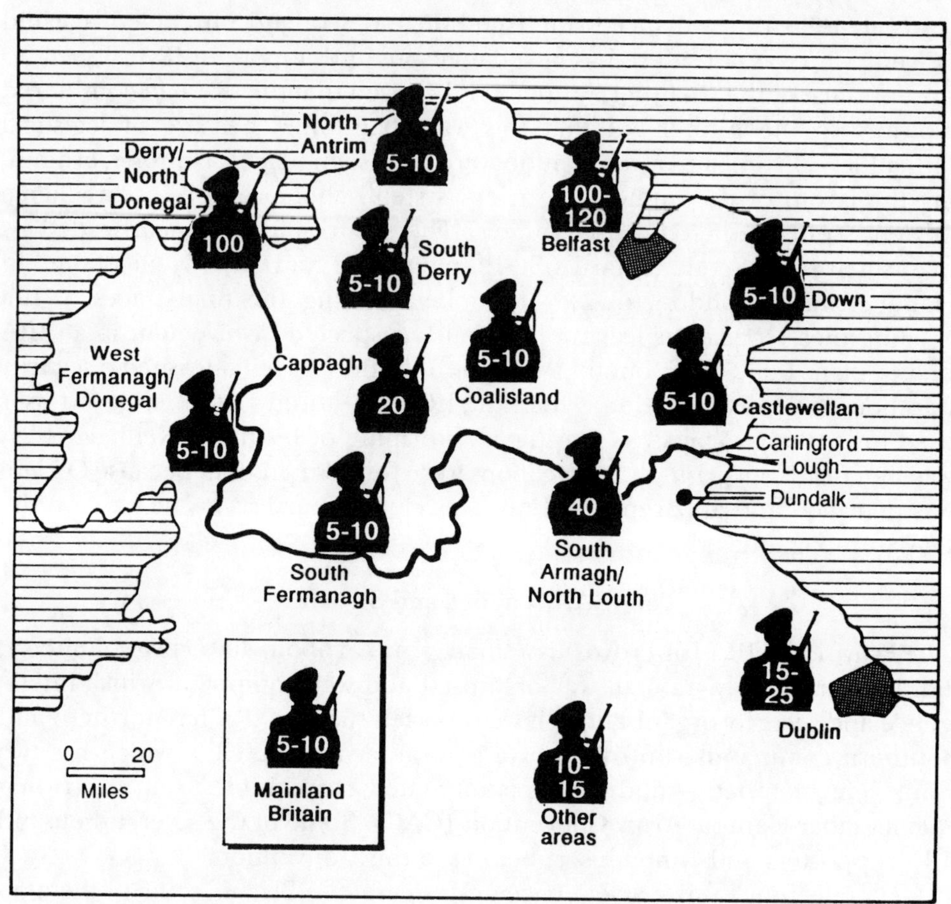

IRA Deployment 1997

The security services believed that at the end of 1996 some of the biggest arms dumps were in the Munster area, in the south of the Republic of Ireland.

The IRA's supreme authority, as has been said, is the GAC and its principal function is to elect a 12-person executive. In turn, the executive chooses a new army council and a chief of staff. the GAC, which meets infrequently, is thought to be the only body with the power to declare permanent peace. It met in 1986 and the next meeting was called for the week ending 15 September 1996. The media was quick to label it 'the terror summit'. British intelligence sources surmised that the meeting had been forced by the rank and file who were dissatisfied about what they regarded as the low level of IRA activity.

The GAC is made up of delegates from every IRA unit, each 'battalion', 'brigade' and 'command', as well as general headquarters, the army council and the executive committee. It exists to give everybody a say in planning, even if the actual decisions are made by only a few people.[7]

IRA Successes and Failures

Despite the feeling among many IRA terrorists and the various splinter groups that the terrorism momentum was inadequate in 1996 a part list of 'incidents' between 9 February and 15 July 1996 shows that there was little respite for the security services:

9 February: The South Quay (Canary Wharf) truck bomb.
15 February: A 5lb Semtex bomb was left in a hold-all in a telephone booth in Charing Cross Road, in central London.
18 February: Edward O'Brien, an IRA man, was killed by his own bomb in a London bus.
9 March: A bomb exploded in Old Brompton Road, West London.
17 April: An empty house was blown up in The Boltons, an exclusive square near Earl's Court, London.
24 April: Two devices were placed at the south side of Hammersmith Bridge, London. The detonators went off but failed to fire the bomb. Scotland Yard stated that the twin devices represented the largest high-explosive bomb planted in mainland Britain.
15 June: A huge bomb exploded in Manchester city centre, injuring 200 people and causing hundreds of millions of pounds damage.
28 June: Terrorists fired three mortars at the army's Quebec barracks in Osnabruck, Germany.
7 June: In the Irish Republic, police detective Jerry McCabe was shot dead and his partner wounded during an attempt by an IRA gang to rob a post office van the policemen were escorting in County Limerick.
13 July: A 1,200lb car bomb destroyed the Kilyhelvin Hotel, Enniskillen, Northern Ireland.
12 July: Police recovered components for about 36 bombs and arrested seven men in South London and a man and woman in Birmingham.

One of the IRA's most devastating attacks took place on 7 October 1996 when terrorists penetrated several hundred yards into the army's headquarters at Lisburn, County Antrim, and planted two massive car bombs at Thiepval Barracks. The first went off in a car park, the second outside a medical centre. It was at once clear that the second and later blast was intended to catch people attending to the casualties of the first explosion. One soldier was killed and 37 were wounded, several seriously.

The terrorist attack on what should have been one of the most secure areas of Northern Ireland was a catastrophic breach of security. The base is protected by physical defences of various kinds and guarded by armed soldiers. Closed circuit television security cameras monitor all entrances and there was a rule at the time of the explosions that all vehicles, even military ones, must be stopped and searched. For some time before the attack intelligence sources had been warning of the probability of a major terrorist attack – and this made the security breach all the more serious.

It was unclear at the time which of the republican terrorist groups had been responsible, the IRA itself, the Irish National Liberation Army (INLA) or the Continuity Army Council of the IRA (CAC), a dissident group which favours ceaseless terrorism with the object of driving the British out of the province. A week before the Lisburn attack the CAC had planted a 250lb car bomb in Belfast city centre; it was defused after a telephoned warning to a local newspaper.

Incidents proliferated in December 1996 and January 1997. On 19 December a well-known North Belfast republican, Eddie Copeland, was injured when a boobytrap went off under his car. A similar device was attached to a vehicle belonging to a former republican prisoner in Londonderry but it was seen and dealt with. The security services blamed the Protestant paramilitary group, the Ulster Volunteer Force.

On 21 December a prominent loyalist politician in Northern Ireland, Nigel Dodds, and his wife were visiting their baby who was seriously ill in hospital in West Belfast when two armed men wearing dark wigs gained entry to the hospital. A police guard recognised one of the men as a well-known member of the IRA. The two terrorists then opened fire. A bullet wounded one policeman in the foot but Mr Dodds, the intended target, was unharmed. Dodds is a close associate of the Reverend Ian Paisley, leader of the Democratic Unionist Party 'I don't know how low these people can stoop', Paisley said. 'It proves conclusively what I have been saying for a very long time – you cannot negotiate with such people. They are beyond the pale.'

On New Year's Eve IRA terrorists placed a 1,000lb bomb in two rubbish bins inside a van, which they left in the grounds of the Belfast Castle, a well-known hotel. A member of the public became suspicious of the van and reported it to the police. This led to a three-day military operation to defuse the massive bomb. The Royal Ulster Constabulary and the army had no doubt that the IRA's intention was to lure the security services to the spot and then detonate the bomb. It would have wrecked even an armoured vehicle, killing or maiming the crew. Even if this had not happened, the Belfast Castle is a popular venue, especially with teenagers and several functions were taking place there on New Year's Eve. Not surprisingly, RUC Chief Superintendent William Davidson said that the IRA was seeking to commit mass murder.

Leaders' Refusal to Condemn Violence

The rest of the world might consider the terrorists' activities as extreme and intolerable but the thinking of IRA activists is that an organisation dedicated to

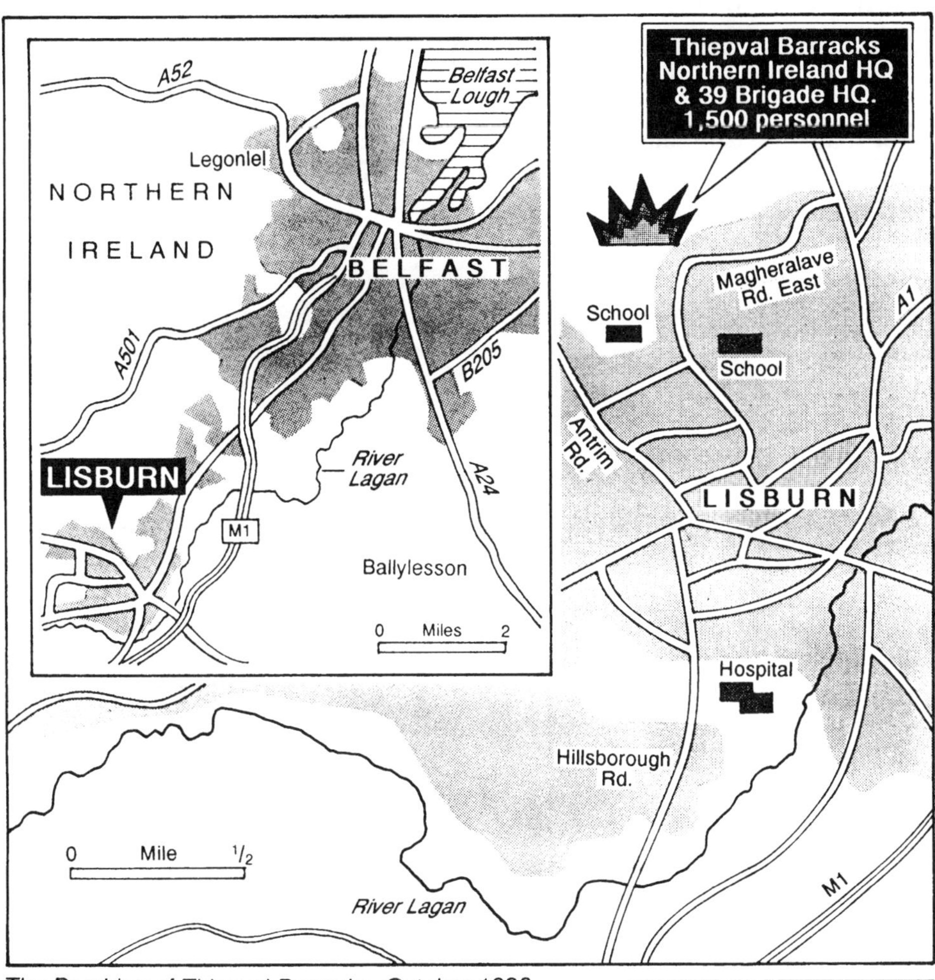

The Bombing of Thiepval Barracks, October 1996

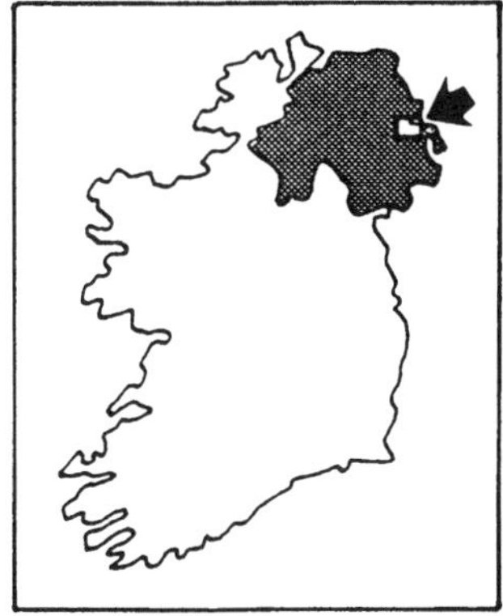

violence must be seen to be violent.

The leaders of Sinn Fein, Gerry Adams and Martin McGuiness, have refused to condemn violence, though they say they 'regret' the loss of life and the injuries to people. According to Sinn Fein, the casualties, however, are the fault of the British government. Simultaneously, Sinn Fein and its supporters pursue a propaganda campaign to convince foreigners, especially in the US, that IRA activities are not aimed at 'ordinary people', only at the apparatus of the state, including the security services.

But humanity is entirely absent in Sinn Fein/IRA. Their terrorist campaigns, especially in Britain, are exercises in manipulation. Society must be afflicted with disorder, apprehension, panic and worry.[8] A sense of outrage must be expressed in public statements in the media, and from the pulpit because public outrage might induce the government to impose restrictions on freedom of movement. For instance, the Cabinet might even reintroduce internment for men known to be terrorists but who cannot be brought to trial for lack of evidence. This would enrage the republican community of Northern Ireland and the government of Eire, which would suit Sinn Fein/IRA very well.

After the Docklands bombing – and every other major terrorist outrage – anger in Britain engendered suggestions that the SAS be ordered to seek out all identifiable IRA terrorists and eliminate them. The terrorist masters would welcome such an extreme step because they know it would be condemned by the international community and that it would prove what they have always stated – that Northern Ireland is governed 'oppressively'. After the London bombing, a senior Conservative suggested that the IRA was engaged in a 20-year strategy of 'incremental terrorism'. By this he meant that the IRA would agree to ceasefires to allow republican demands to be moved forward in a political way but each ceasefire would be followed by a return to violence, and so on indefinitely, until the British government of the day abandoned Northern Ireland altogether.

The Protestant Loyalist (Unionist) paramilitary/terrorist groups declared their own cease fire soon after the IRA ceasefire came into effect. This meant that attacks on known or suspected IRA members and their supporters stopped. The leaders always said that their ceasefire was conditional on the IRA holding to its suspension of operational activities but even after the breaking of the ceasefire with the Docklands bombing the Loyalists maintained their position. However, it came to an end in the final days of 1996. The Ulster Defence Association and other groups said that they had been 'goaded' into retaliation against IRA terrorist acts.

Also at an end were any benefits that had resulted from the elections. International observers said that the poll had been democratic and fair – at least by the devious standard of Northern Ireland – but discord ruled from the moment that the elected representatives met under the chairmanship of the American former senator, George Mitchell. While he did his best, the various parties could find no common ground and the hatreds and distrust of centuries prevented any real progress.

The new British Labour government of Tony Blair hoped to make progress where John Major's Conservatives had failed. Indeed, on 16 June 1997 Blair's Northern Ireland Secretary, Dr Mo Mowlam, was due to hold talks about defusing sectarian conflict over a series of Protestant matches. The next day IRA gunmen shot dead two Northern Ireland policemen. The organisation made no secret of its operation. Spokesmen for the US and the Republic of Ireland condemned the murders. Ms Mowlam said, 'It is a sad day for trying to move peace forward', and urged Loyalist gunmen not to retaliate.

Ms Mowlam took a courageous step on 7 August when she met Gerry Adams and other Nationalist leaders openly in yet a further attempt to reach some degree of trust. Afterwards both Adams and Mowlam paid compliments to the other but veteran observers warned against expecting too much from Mowlam's sincere and very ambitious hopes.

References

1. *The Sunday Times*, 23 June 1996.
2. Speaking to a press conference, 17 June 1996.
3. *Canberra Times*, Australia, 24 February 1996.
4. Documents found in O'Brien's home revealed, under scrutiny by MI5, that the IRA had planned a major attack on the Houses of Parliament. It involved mooring a barge packed with two tons of explosive on the Thames opposite the Palace of Westminster. It would have been timed to explode when up to 10,000 people were in the Houses of Parliament. The blast would have destroyed much of St Thomas' Hospital on the opposite bank of the Thames. Parliament had been an IRA target in 1974 when a bomb damaged part of Westminster Hall; and in 1979 a car bomb in the courtyard of the House of Commons killed Airey Neave, a spokesman on Northern Ireland and a close friend of Lady Thatcher.
5. On 12 July 1692 William of Orange defeated his Catholic rival, King James II, at the Battle of the Boyne. This victory established England's Protestant ascendancy in Ireland. Ever since that time each year Protestants belonging to the Orange Order don bowler hats and orange sash-collars and make a series of marches to commemorate the Battle of the Boyne. And each year the Catholics react with fury.
6. In 1975, a young police constable, Stephen Tibble, tried to apprehend an IRA suspect leaving a terrorist safe house in Hammersmith. Constable Tibble was shot dead. This incident had a profound and enduring effect on all ranks at the Metropolitan Police and naturally made them ultra-cautious when approaching IRA suspects.
7. For people accustomed to associating the terms battalion, brigade, command and general headquarters with normal armies it is confusing to see them used by the IRA for small numbers of men and women. A British army battalion used to number 1,000 and even in modern times it is 600; an IRA battalion might consist of no more than 30. A regular brigade could have 3,000 members; an IRA brigade probably has about 60 plus 'non-operational supporters'. No fixed establishment of numbers exists in the IRA. The grandiose term 'Irish Republican Army' is intended to indicate that this 'people's army' runs into scores of thousands, just as a regular army does. The use of the terms for small units – brigade, battalion and so forth – has another purpose: the IRA sees them as justification for its claims to be a legitimate military formation whose members, when captured, should be treated as prisoners of war rather than as common criminals. In all their propaganda, Sinn Fein/IRA refers to its

gaoled members as political prisoners.

8. 'Only when deeds as well as words show that violence is fruitless and its advocates unheard will Ulster enjoy real peace.' *The Times*, London, 24 September 1996.

24

Refugees: The Result and Cause of Wars

DISASTERS IN THE COMMONWEALTH OF INDEPENDENT STATES

The refugee disasters of Africa, such as those of Rwanda-Burundi, and of former Yugoslavia, have been well publicised and thoroughly studied. The armed conflicts within the countries of the Commonwealth of Independent States (CIS) and the refugee movements that have so been caused have, in contrast, received relatively little media coverage. One reason could be the general ignorance about the CIS and Central Asia, where they are located. Another could be that these countries are less accessible than Central Africa and Bosnia-Herzegovina and Croatia and the other slavic states.

However, as the millennium comes to an end it is appropriate to draw attention to the human cost of the wars and 'civil unrest' in the CIS, if only because the bitterness and anger among millions of refugees is a trigger for further wars. Wars have raged in the CIS every day since the break-up of the Soviet Union, just as I and other observers warned that they would.

Certainly, some of the migratory movements result from civil and ecological disasters – such as the nuclear catastrophe at Chernobyl in 1986 and the total pollution of the Aral Sea since then – but political/military upheavals account for most of the nine million refugees living in misery in the CIS in 1996–97. These migratory movements continue a trend established by the ethnic barbarity of Stalin, who was responsible for more refugees than can be accurately estimated but they included three million Soviet citizens from 20 ethnic groups. He deported en masse Volga Germans, Kalmyks, Chechens, Ingush, Karachavs, Balkars, Crimean Tatars and Meshkhetians.

Below I record the main refugee movements of the 1990s, resulting from war, 'ethnic cleansing', political pressure and intimidation. They were discussed at a conference in Geneva in May 1996, organised by the United Nations High Commission for Refugees (UNHCR), the International Organisation for Migration (IOM) and the Organisation for Security and Co-operation in Europe (OSCE).[1]

Russia

After the break up of the Soviet Union, 2.7 million people, mostly ethnic Russians, migrated to Russia from the former republics. The greatest exodus was from Kazakhstan, Tajikistan and Uzbekistan, countries affected by Islamic fundamentalism and its consequent violence. Many others came from Georgia. Conflicts in Ingushetia, North Ossetia and Chechnya produced many thousands of 'internally displaced' people. Large numbers of Armenians, afraid of further conflict with Azerbaijan, have entered Russia, as have deported Crimean Tatars and Germans. About one million Russians from the Far North and Far East migrated west in the period 1990–94.

Ingushetia–North Ossetia

In 1991 a law was passed providing for the 'rehabilitation of the rights of deported peoples'. Controversially, it included the right to regain territory seized from deportees by Stalin. The dictator, having deported the Ingush people, transferred the Prigorodnyy region of Ingushetia to North Ossetia. When the Ingushi returned, expecting to have this land given back to them, a war broke out. More than 500 people were killed and 3,000 wounded while the North Ossetians 'ethnically cleansed' up to 75,000 Ingushi from the area. General Dzohokar Dudayev, unilaterally declared Chechnya an independent republic in 1991 appointing himself Prime Minister and Minister of Defence. One result was that Ingushetia was officially established as a separate republic but it still has no recognised demarcated borders.

Chechnya *(See also, Chechnya the Great Russian Humiliation p. 80)*

Dudayev's government and Chechnya's opposition fought an internal war which, between June 1993 and December 1994, created 200,000 refugees and caused many deaths. Russia intervened militarily in December 1994 and carnage and destruction on a massive scale followed. By the end of 1996 the UN estimated that 40,000 people had been killed and 500,000 had been displaced; this is half the country's population. About 180,000 Chechnyans have gone to other parts of the North Caucasus, principally Ingushetia (50,000), Dagestan (30,000) and Stavropol (25,000).

Belarus

Belarus has such an extraordinary mixture of refugee peoples that reliable figures are available only for the period 1993–96. Thirty thousand sought refugee status, 84 per cent of them Russian-speaking. They came from Latvia, Russia, Lithuania, Georgia, Kazakhstan, Tajikistan and Estonia. Others came from Afghanistan, Ethiopia, Iran, Vietnam and Somalia. The Belarus government claims that at least 100,000 refugees are illegally resident in the country.

Central Asian States – Potential Battlegrounds

Moldova

This country is like a great refugee transit camp, with Afghans, Bangladeshis, Indians, Jordanians, Nigerians, Pakistanis, Sri Lankans and Syrians passing through to other destinations. Nevertheless, many stay for considerable periods.

Ukraine

Ukraine's economy and social cohesion has been destabilised by a massive refugee problem. Between 1992 and 1994 more than a million refugees arrived there from Moldova and the Baltic states, almost half of them ethnic Ukrainians. In the same period 738,000 people left Ukraine, of whom 226,000 were Ukrainian. The following year 270,000 left Ukraine and 178,000 arrived. Asylum seekers continue to arrive from Somalia, Ethiopia, Iran, Sri Lanka, Afghanistan, Moldova and the Chechen Republic. Many Bulgarians, Armeninians, Greeks and Germans are also resident in Ukraine.

Armenia *(See also Armenia and Nagorna-Karabakh p. 60)*

Large-scale emigration has taken place to Russia but Armenia has 310,000 refugees as a result of the conflict in Nagorny Karabakh. Azeri bombardment of border areas made 72,000 homeless and about one million people live in temporary accommodation. Many Armenian families left all their possessions in Azerbaijan when forced out by the Azeris. Many other refugees fled from ethnic violence between Georgians and Abkhazians.

Azerbaijan

Following the conflict over Nagorny Karabakh, more than 700,000 'internally displaced persons' live in Azerbaijan. In addition, there are 300,000 refugees. They include 185,000 ethnic Azeris from Armenia, 48,000 Meshkhetians and 5,000 people from the Russian Federation.

Georgia

Georgia has more than 300,000 refugees and displaced persons. Most of them are the result of wars in South Ossetia and Abkhazia and of civil unrest elsewhere in the Caucasus, notably North Ossetia, Ingushetia and Chechnya. Smaller numbers of people from Meshkhetia, Azerbaijan and Russia have also found refuge there.

South Ossetia

In 1989 South Ossetians began a war against the Georgian authorities because they wanted to unify South Ossetia with the North Ossetian Republic, a member of the Russian Federation. The result of this conflict was that many South Ossetians sought refuge in North Ossetia while many Georgians, now vulnerable in South Ossetia, left that region to find sanctuary in other parts of Georgia. In July 1992 a peacekeeping force was stationed in South Ossetia and the Organisation for Security and Co-operation in Europe established a mission there. To further the peace process, in May 1996 Russia, Georgia, North Ossetia and South Ossetia signed a memorandum (not a treaty) rejecting the use of force and the persecution of ethnic groups. Demilitarised zones were also created. Conditions have greatly improved but no reverse refugee trend has taken place.

Abkhazia

The Abkhaz people want to be independent from Georgia, whose government fiercely resists their attempts to join the Russian Federation. In August 1992 Georgian troops entered the Abkhaz capital, Sukhumi, and were resisted by the Abkhaz Interior Ministry Forces. This war caused a remarkable reciprocal refugee movement: about 275,000 ethnic Georgians fled to eastern Georgia while many Abkhaz in other parts of Georgia fled to Abkhazia. A ceasefire was arranged in late 1993 and some Georgians then returned to Abkhaz and their former homes but all those suspected of taking part in the fighting were barred.

The ceasefire is frequently broken and many more people are killed or wounded than the authorities acknowledge.

Kazakhstan

In the early 1940s about one million Germans were resident in Kazakhstan, deported there by Hitler to build up the Nazis' claims to other lands. Since 1990–91 more than 600,000 of these Germans have left, mostly for Germany. After 1991, 125,000 ethnic Kazakhs came home from Mongolia, Afghanistan, Tajikistan, Turkmenistan, Kyrgyzstan, lured by promises of financial support. Militant Kazakh groups not infrequently attack 'foreigners', mostly those of Caucasian origin.

Kyrgyzstan

Following the Second World War, many Russians and Ukrainians fled to Kyrgyzstan to escape Communist pogroms. People from East Germany (as it was until 1991) fled from that country to escape from economic poverty. Since 1989 about ten per cent of the one million Russians and 110,000 Germans and others have left the country. In southern Kyrgyzstan violence is endemic between Kyrgyz and Uzbeks, mostly over land rights. The UN estimates that three thousand people have been killed. Many Uygurs, a people related to the Uzbeks, have crossed into Kyrgyzstan from Xinjiang, China. Within Kyrgyzstan another significant trend is taking place – the displacement of tens of thousands of peasants who want to settle in the capital, Bishkek.

Tajikistan

Tajikistan is an unhappy but classic example of what can happen as a result of a civil war. Breaking out in 1992, this conflict has created an estimated 900,000 refugees and internally displaced people. Tens of thousands fled for their lives to other countries within the CIS and even to war-torn Afghanistan. When a ceasefire was agreed in 1994 thousands returned but 200,000 Tajiks were unlikely to return from their CIS 'havens'. While Tajiks went to Afghanistan, probably 1,000 Afghans have fled to Tajikistan. Non-Tajik ethnic groups such as Germans, Russians and Crimean Tatars, are emigrating, mainly to Russia.

Tajikistan is in a parlous state. The fighting between government forces and rebels merely intensified after another ceasefire in July 1996 that was supposed to end the five-year-war. The deputy leader of the Tajik Islamic opposition, Daviat Usmond, said on 13 August that because of government attacks on his positions the ceasefire agreement was not observed 'even for an hour'.

Disturbingly, the rebels refer to themselves as the mujahideen or holy war warriors, as do the *Hezbollah* fighters of Lebanon and the tribes of Afghanistan during their war against the Soviet Union. The label implies extremism. The rebels have been successful in some areas and in August 1996 they captured the town of Tavildara in central Tajikstan. Soon after this, the chief of the Russian Federal Border Service – a quasi-military organisation – warned that Russian border guards stationed in Tajikistan might resume 'preventive attacks' on Tajik rebel positions in Afghanistan.

Soon after this threat General Pavel Tarasenko, commanding the Border Service, accused the Afghan government of providing helicopters to the mujahideen rebels in Tajikistan. All these events were to some extent overshadowed by the Taliban's defeat of the Afghanistan government forces in September and October 1996. However, the Russians are intent on propping up the weak ex-communist states, preferring this course to any form of Islamic militancy. The chances of peace in Tajikistan are slight. It is the poorest of the former Soviet Central Asian states and by early 1997 the war was estimated to have killed 40,000 people and created 800,000 refugees.

Turkmenistan

Ethnic Turkmens are returning to Turkmenistan from various parts of Central Asia, many of them from Russia, while Russian settlers in Turkmenistan are anxious to return to Russia.

Uzbekistan

In 1989 the Uzbeks and Meshkhetians turned on each other and the Meshkhetians suffered grievously. Nearly 200,000 became homeless and the Russian army evacuated many from the Fergana Valley in order to save their lives. Most fled from Uzbekistan and 45,000 went to Azerbaijan. Following the dissolution of the Soviet Union many more Meshkhetians, fearing massacres, fled from Uzbekistan. Afghans and Tajiks fled from wars in their own countries and settled in Uzbekistan. Many of the ethnic Russians have abandoned Uzbekistan and the majority of the Jewish people, long settled in Central Asia, have departed following religious intolerance.

References

1. A publication of UNHCR in May 1996 stated: 'Since 1989 around 9 million people (excluding military transfers and voluntary migration) have moved within or between the countries of CIS – one in every 30 of the region's inhabitants. The movements are the largest, most complex and potentially the most destabilising to have taken place in any single region since the end of World War II.' This is, in effect, a warning of detonators that could explode into further wars. Much of this summary was published in a British Foreign and Commonwealth Office 'Background Brief' in July 1996.

25

Somalia Civil War

When Major General Siyad Barre seized power in a coup in 1990 he set Somalia on the path to ruin. He fought a disastrous war against Ethiopia and ruthlessly crushed all opposition at home until the Somalia National Movement (SNM) rose against him. The offensive which deposed him began on 30 December 1990. The SNM joined with the recently-formed United Somali Congress (USC) and the Somali Patriotic Movement (SPM) and combined in all-out assault.

The SNM, dominated by the Isaq tribe, approached Mogadishu, the capital, from the north while the SPM of the Ogadeni tribe swept up from the south. Desperately, Barre tried intimidation, bribery and promises, including a new constitution and a referendum which would, he assured the people, lead to a multiparty democracy. Nothing worked and the rebels and government forces fought a no-quarter-given battle that lasted a full month in January 1992.

Barre, having presided over the destruction of his country, during which perhaps 500,000 people died (out of a population of 5.2 million), fled to Kenya. But there was no peace for the survivors of his regime because a new phase of the civil war began. The large Hawiye clan split into two main factions, one following the interim president of Somalia, Ali Mahdi, the other loyal to General Muhammad Farrah Aidid's faction. From December 1991 the conflict was so ferocious that the US Security Council called on both sides to respect a ceasefire so that humanitarian aid could be sent into the country but the warring groups were more interested in importing arms than food.

Neither side to the war had any sense of strategy; their only objective was to kill and destroy and since AK-47s and M-16s abounded, with limitless ammunition, Somalia's predicament was calamitous. Apart from the tens of thousands of deaths, about 250,000 people had been driven from their homes and 70 per cent of the population were suffering from malnutrition. The risk of death was ever-present because hundreds of gangs of young fighters, driving pick-up trucks with 50mm machine-guns mounted on the back, roamed the streets spraying bullets wildly.

With great reluctance, the UN sent in 30,000 peacekeepers in 'Operation Restore Hope'. The largest contingent was from the US, with Pakistani,

Nigerian and Australian support. American marines landed near Mogadishu in December 1992 and embarked on a campaign to kill or capture General Muhammad Aidid, the principal warlord. The use of helicopter gunships was not only fruitless but foolish because inevitably Somali citizens were killed.

Among the US marines who took part in Operation Restore Hope was Corporal Hussein Aidid, son of General Aidid. It is not clear if this was a mistaken posting by officials who did not know of his background or a deliberate attempt to use Hussein as a human shield, since General Aidid's fighters would not know if they were spraying bullets at the son of their own leader. When the US issued a warrant for his father's arrest, Hussein was returned to the US.

In June 1993 Aidid's fighters killed 24 Pakistani soldiers and on 3 October 18 Americans. On 31 March 1994 the Americans pulled out of Somalia in 'Operation Quickdraw'. Naturally, President Clinton and his secretaries of state tried to make the departure from Somalia seem like a withdrawal rather than a retreat but American prestige suffered. The US Land Force Commander, Major General Thomas Montgomery, announced that eleventh-hour declarations, signed by the two chief warlords under pressure from the UN, would 'save Somalia from continuing anarchy'. It was a face-saving measure and it had no chance of working. With other factions becoming organised, Somalia faced a future of continuous tension and uncertainty.

The War in 1996

The principal warlords and claimants for Somalia's presidency were General Aidid, his reputation greatly enhanced by the American withdrawal and consequent humiliation, Ali Mahdi Muhammad and Ali Hassan Osman, whose *nom de guerre* is 'Otto' or 'Ato'. Mahdi and Otto shared close clan sub-ties through membership of the Ayer clan but even so, had serious disagreements. In 1994 Otto was Aidid's financier. Mahdi did not like the Otto–Aidid axis and managed to break it by giving non-military aid to Otto in order to further his own ambitions. All three men were from the great Hawiye clan, but Ali Mahdi came from the clan-family Abgal and the other two from the Habre Cedir clan-family. The complex family and clan relationships affect everything in Somalia but there is yet another factor, the 'blood compensation', which is the Somali form of vendetta or the Arab 'eye for an eye'. Many deaths occur because of this tradition.

In April and May 1996 Aidid's fighters and those of Ali Otto fought ferocious village and street battles, mostly in south Mogadishu and Merca port. About 300 people were killed and thousands were displaced. Aidid had assumed that because his supporters had bestowed the title 'president' upon him in June 1995 he was indeed the nation's legitimate leader, but the only country to recognise him as such was Libya – and President Gaddafi's approval counted for little. To enlarge and protect his 'presidency' Aidid withdrew many of his militiamen from the Bay Region, of which Baidoa is the centre, to strengthen

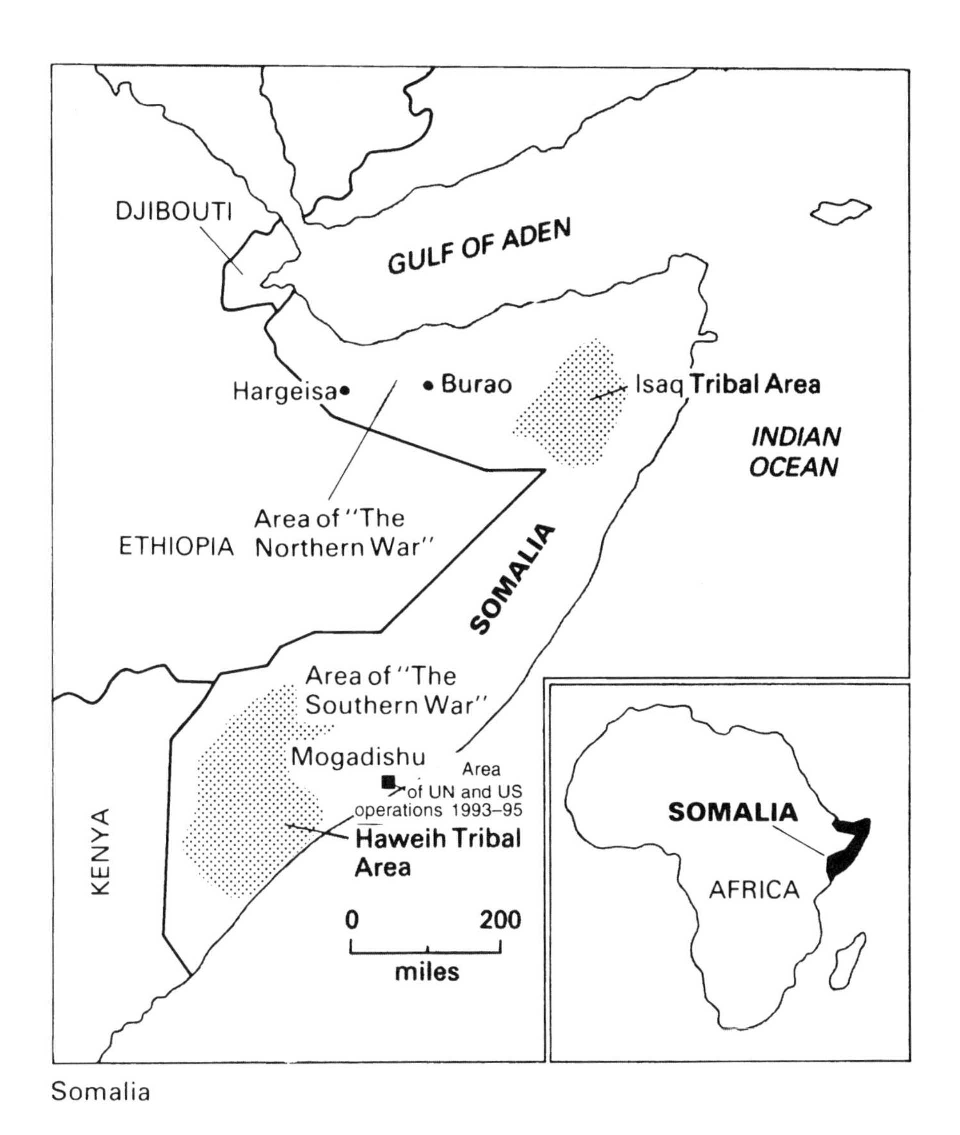
DJIBOUTI
GULF OF ADEN
Hargeisa
Burao
Isaq Tribal Area
INDIAN OCEAN
Area of "The Northern War"
ETHIOPIA
SOMALIA
Area of "The Southern War"
Mogadishu
Area of UN and US operations 1993–95
Haweih Tribal Area
KENYA
0
200
miles
SOMALIA
AFRICA

Somalia

his hold on southern Mogadishu. This was a retrograde step because he had captured the Bay region in September 1995 and now he was giving it up. His next setback came from the Rahanwein Resistance Army (RRA) which captured Hoddur, the main town in the Bakool Region.

Diplomats from neighbouring Kenya, who are better able to move freely in Somalia than other foreigners, said in June 1996 that Aidid's personal militia consisted of about 1,000 men, most of them veterans of conflict. They had modern vehicles, the equivalent of Land Rovers, armed with heavy machine-guns. The Kenyans also reported that Aidid's deputy, Abderrahman Ahmed Ali Tur, often visited Libya to ask Gaddafi for more money and equipment. Even so, the Kenyans added, the fighters were not paid regularly and reverses had caused some of them to defect.

Aidid's main source of revenue was from the sale of bananas and qat, the narcotic plant. These commodities are shipped through Merca, which is why Aidid had to hold the port. It was estimated that through bananas, qat, blackmail and protection money Aidid had an annual revenue of more than $1 million, a considerable amount for a warlord whose costs are minimal. But, alarmingly for Aidid, in October 1995 Ali Mahdi shelled Aidid's positions at Merca. To avoid a costly battle for Merca, Ali Mahdi developed his own new port at Al Maan, north of Mogadishu.

An endless debate among ordinary Somalis during 1995 and 1996 was about the relative merits of Aidid and Ali Mahdi as national leader. Cunningly, Ali Mahdi played the Islamic law card; he backed the establishment of the High *Sharia* Implementation Court, which imposes harsh sentences under Islamic law. The suffering Somalis believed that this offered them some security and protection against the marauding killers and thieves. Allied with Ali Mahdi is Moussa Sudi, a fundamentalist mullah with great influence who is also a personal friend of Libyan President Gaddafi.

General Aidid's Life and Death

On 1 August 1996, General Aidid was hit by two bullets during the fighting in the central Medina district of Mogadishu. He underwent surgery at his home but died from his wounds, complicated by a liver disease. It is not known whether Aidid was aged 59 or 62. That he should die during the fighting was predictable after he had become increasingly active as a frontline commander of his troops. More than 100 fighters were killed and 400 wounded during the battle in which Aidid lost his life.

Aidid, whose name means 'man of steel', was a career army officer, trained in Italy and the Soviet Union. It was an unlikely career for he was the son of a camel herder. On Somalia's independence in 1960, Aidid was included in the newly-formed national army and he rose through the ranks which brought General Siyad Barre to power in 1969. Barre named him as chief of military intelligence, then gaoled him on suspicion of plotting to seize power. He was in prison until 1975. It was said later than he was so hungry that he ate the soap with which he had been issued to wash himself before Friday prayers.

He wanted revenge against Barre and his six years of brooding turned him into an introspective and irrational man. Barre apparently regretted his brutal treatment of Aidid and used him as an advisor during Somalia's invasion of Ethiopia's Ogaden region in 1977. Aidid then accepted the post of Somali ambassador to India, which gave him an opportunity to plot against Barre. Slowly, he built up an opposition movement with its base in Addis Ababa, Ethiopia. General Mengistu of Ethiopia was happy to welcome and support any Somalis operating against their own government. Aidid's United Somali Congress – Somali National Alliance (USC–SNA) was specially welcome. In 1988 all this changed when the Ethiopian and Somali governments became friendly again and the USC-SNA had no option but to return to Somalia. The civil war resulted from this move.

Aidid faced another problem when Siyad Barre fell because the USC-SNA fell apart along tribal lines. The war exacerbated an already serious famine and Aidid was increasingly blamed for his failure to resolve the twin political-military crises which kept the famine going. On 17 June 1993 the UN forces in Somalia issued a warrant for Aidid's arrest when he was accused of ordering the ambush in which 24 Pakistani peacekeepers were killed and their bodies mutilated. Aidid was successful in humiliating the US-led force but instead of capitalising on the prestige this gave him in Somalia he went back to fighting the rival clans and capturing towns, the yardstick by which Somali warlords assess their success.[1]

The New Aidid

Hussein Aidid had left the US army in 1995 and returned home to Somalia to marry. His father then appointed him head of security in Baidoa, where he went about his work in American combat gear with a pistol on his hip. His swaggering style made him a cult figure. General Aidid's followers, while accusing the CIA of arranging his death, saw no inconsistency in appointing Hussein, a former US marine, his successor. Hussein at once declared that his 'government' would continue with his father's 'high ideals' of pacifying Somalia. He meant that he would fight any individual or group who opposed him and his father's followers at once gave him hero status.

In a great ceremony at Mogadishu stadium Hussein Aidid was sworn in as 'interim president', a title that was at once opposed by senior members of the Somalia National Alliance. Also opposing it were the SNA's enemies in north Mogadishu.

Aidid particularly appeals to the *moreyhan* (bandits), the young wild bush fighters recruited by his father. He signed them on when they were in their early teens and gave them licence to do as they pleased. At the age of 31, Hussein is young himself and he promises the *moreyhan* 'glory'.[2]

At the end of 1996 Somalia had disintegrated into violent factions and the war was even more ferocious than during the time of Barre or Aidid, according

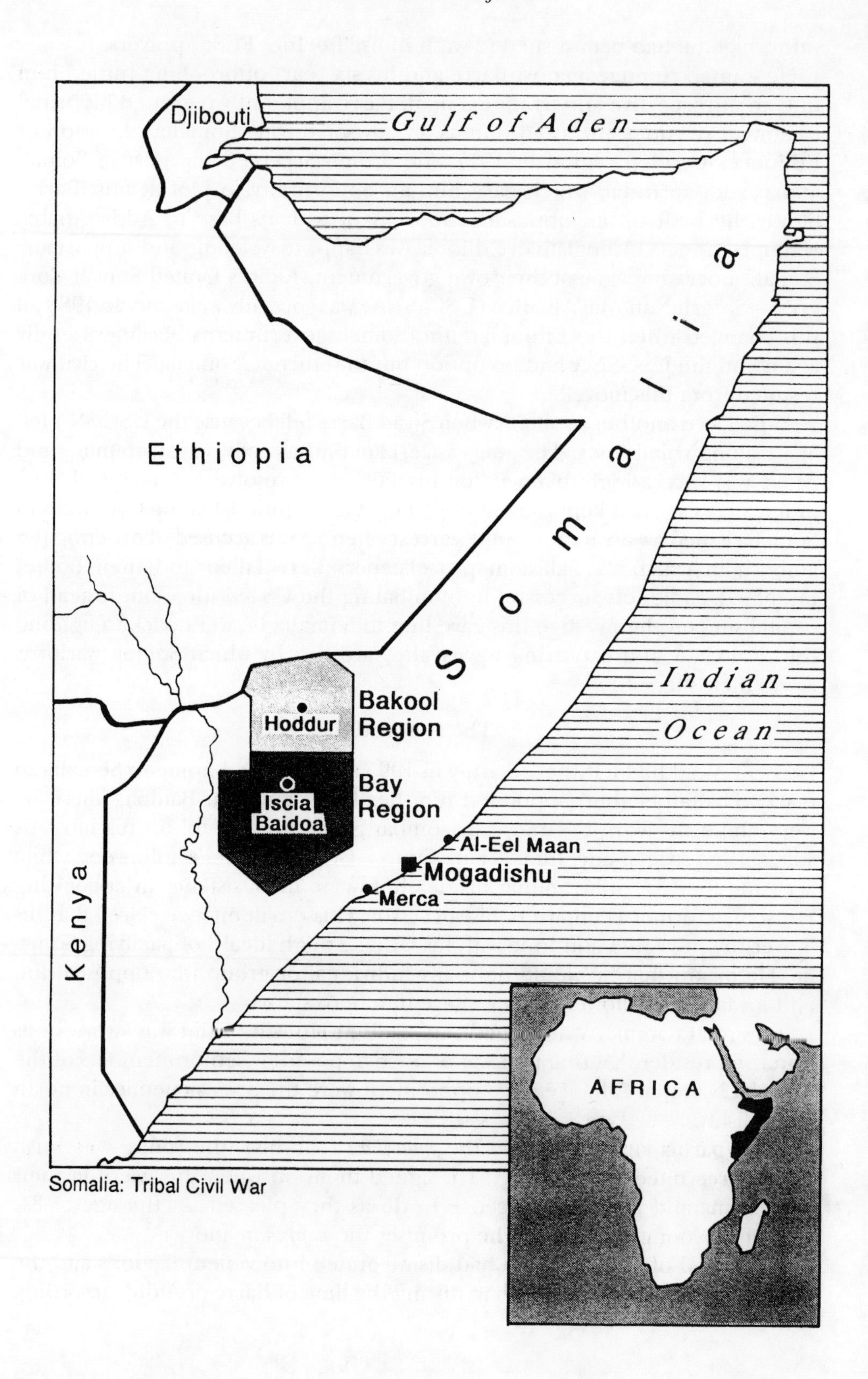

Somalia: Tribal Civil War

to veteran foreign aid workers. Firing was indiscriminate and, as always, the innocent suffered. Late in November, in southern Mogadishu, a mortar shell hit a school, killing eight children and wounding 20 others.

Ethiopian Raids

While Somalia was still coming to terms with its future after General Aidid, a new 'war' broke out between Ethiopia and Somali fundamentalist militia groups. The precipitating cause of the flare-up were the actions by the *al-Ittihad* (Islamic Union) which had established its headquarters in the town of Lugh, 50 miles from the Kenyan border, in 1993. It imposed *Sharia* (Islamic law) in an attempt to end the chaos that had destroyed much of Somalia. The *al-Ittihad* spread its influence to take over Dolo and Belet-Hawa.

But it went much further than this. Allying itself with ethnic Somalis in the Ogaden region of Ethiopia – the scene of intermittent Ethiopian-Somali wars – *al-Ittihad* extremists made leading Ethiopian ministers and officials its target. Bombings and assassination attempts in Addis Ababa were commonplace. These terrorist attacks were made in the name of *jihad*, or holy war, against infidels. Bomb attacks on hotels in Addis Ababa forced the government into retaliatory action. Ethiopian helicopter gunships, tanks and artillery attacked *al-Ittihad's* bases and sanctuaries.

The *al-Ittihad* made extravagant claims of its successes against the Ethiopians: that they had shot down a helicopter, set fire to some tanks, captured army vehicles and killed 200 soldiers. The Ethiopians, who could not have observed the success of their own strikes, made no claims but threatened that further 'international terrorism' emanating from Somalia would result in further punitive action. The reference to international terrorism was deliberate; President Gaddafi of Libya and President al-Bashir of Sudan constantly try to take over Islamic fundamentalist groups which they can then use to their political advantage.

References

1. Aidid's obituary in *The Times*, London, 6 August 1996, states: 'Aidid refused to accept that he had become part of the problem rather than a solution. ... He was always immaculately dressed and his neatly-ironed shirts remained uncreased even as the country he hoped to rule collapsed into war, famine and anarchy.'
2. Sam Kiley, who lives dangerously as *The Times* Africa correspondent, has spent much time in Somalia and knows its society well. He wrote, (*The Times*, 6 August 1996) 'Like child soldiers throughout Africa, the *moreyhan* entered combat before becoming "ethical beings". Now beyond traditional control, they fight for loot and the sheer excitement of battle.'

26

Sri Lanka Civil War

OFFENSIVES AND COUNTER OFFENSIVES

Background to a Merciless Conflict

The Sri Lankan conflict between the Hindu Tamils and the Buddhist Sinhalese is similar in its fixed and ferocious intensity to that between extremist Protestants and Catholics in Northern Ireland, the hostility between fundamentalist Muslims and zealot Jews on the West Bank, and the tribal hatreds of Hutus and Tutsis in Central Africa. They all have ancient roots, they are fought with a ferocity that prevents compromise and they are apparently insoluble.

The Sri Lanka war has always seemed strange to me because I have found Hindu and Buddhist Sri Lankans pleasant, intelligent and hospitable people. Perhaps this really is their natural mien and the one they show strangers who visit their agreeable land. Yet they are capable of remorseless killing and diabolical cruelty, as wanton as that of machete-wielding Hutus and Tutsis.

The present problems began in 1983 when the Tamils, a minority of 2.5 million people, began a serious campaign to gain a wholly separate and independent state in northern Sri Lanka. The majority Sinhalese, numbering 16 million, rejected the idea of Tamil independence. The Tamils, impatient for results, created several guerrilla groups. In 1983 the best known were the Liberation Tigers of Tamil Eelam (LTTE) and the Tamil Eelam Liberation Organisation (TELO). The guerrillas ambushed and massacred an army patrol, thus ensuring a bloody reprisal. In Colombo, the capital, Sinhalese mobs slaughtered a thousand Tamils.

The Tigers, led by the ruthless and intelligent Vilupillai Prabhakaran, killed all the TELO leaders to ensure that the LTTE were the undisputed masters of what they called the 'liberation movement'. Finance was no problem; it came from the wealthy Indian Tamils and from large Tamil communities worldwide. In Jaffna peninsula, the Tigers soon developed a formal army structure while maintaining a guerrilla wing. The Tamil experiment of combining regular army tactics with guerrilla actions could well be the most efficient in the world.

Sri Lanka's president, J.R. Jayawardene, pleaded with India's prime minister,

Rajiv Gandhi, for military help and Gandhi sent an Indian Peacekeeping Force (IPKF). The operation was a total failure and, having lost 2,500 soldiers the IPKF was withdrawn.

Further emboldened, the LTTE absorbed, took over or crushed all other Tamil organisations and by the end of 1989 was the one Tamil power. In virtually every encounter with the Sri Lankan army the Tigers were triumphant and on 31 December 1990 Prabhakaran declared a ceasefire. The new prime minister, Ranasinghe Premadasa, misreading the situation and trying to exploit it for his own ends, announced that 'terrorism was being defeated'. The ceasefire lasted only two days as the Tigers showed that terrorism was far from defeated. They assassinated the minister of defence, Ranjan Wijeratne, and attacked numerous army posts and bases.

At the end of April 1991, the security forces launched one of its few real offensives, striking at rebels on Karaituvu and Kayts islands. They caused the Tamils 300 casualties, but the Tamil fighters instantly responded, killing 60 soldiers in an ambush.

A new defence minister, Air Chief Marshal Walter Fernando, desperate for a government success, introduced 'Operation Thunderbolt'. At its root was a frequent and random security check of vehicles entering Colombo. In this way, it was hoped, explosives and weapons would be found. The Tigers' response was to use a suicide bomber to explode a car bomb close to Colombo's main military base; it killed 70 people and wounded another 200. Operation Thunderbolt came to an end.

In July 1991 the Tamils attacked army positions at Elephant Pass, the isthmus linking the Sri Lankan mainland to the Jaffna peninsula. It was followed by a major army offensive. These battles continued intermittently for two years. (See *War Annuals 5* and *6*.) When the Tigers assassinated Prime Minister Premedasa diplomats in Colombo in 1994 gloomily predicted to me that the war would last until the end of the century.

The Tigers seemed to be unbeatable but many of their people had left the Jaffna Peninsula and the 600,000 civilians remaining there were short of food. Yet another Prime Minister, Mrs Chandrika Kumaratunga,[1] tried to negotiate with Vilupillai Prabhakaran through the Red Cross and through four peace negotiators who flew into Tamil territory.

In the last quarter of 1994 Colombo seemed less tense than it had for several years and a presidential election campaign was taking place. At a United National Party rally, a Tamil suicide bomber detonated explosives wrapped around her body and thousands of ball bearings ripped like shrapnel through the crowd, killing 55 people and wounding another 200.

It was by now clear that the LTTE was intent on wiping out the entire political leadership of all parties. It was murdering for political gain. Nevertheless, Mrs Kumaratunga having become President, vowed to 'pursue peace'. Despite army opposition, she called for a ceasefire and talks with the LTTE. To prove her goodwill, she eased the embargo on the Jaffna Peninsula and promised to raise

$800 million in foreign aid for reconstruction there.

Again the Tamils broke a brief truce. They sank two more naval patrol craft, shot down two air force planes with missiles nobody knew they possessed, overran an army camp and butchered their prisoners. In two months – February–March 1995 – the LTTE killed more than 1,000 people in a series of attacks that terrified much of Sri Lankan society.

The security forces had already organised Operation LEAP FORWARD and, provoked beyond measure, Mrs Kumaratunga was compelled to implement it, on 9 July 1995. Its objectives were to kill as many guerrillas as possible and to wrest Jaffna Peninsula from the Tigers' control. The army claimed that it was a success but the LTTE's counter-offensive, Operation TIGER LEAP, quickly demonstrated that Prabhakaren's generalship was superior to that of the army chiefs. He allowed the army to move deeply into enemy territory and when the troops were thinly spread he hit back. Cutting army supply lines, his fighters isolated small army units and wiped them out. On 16 July Tigers infiltrated Kankesanturai naval base, where a suicide diver sank the warship *Edithare.*

On 28 July women Tigers, known as 'Freedom Birds', led an attack on army camps in the north-east. Insisting that its forces had retreated in only a few areas, the government said that 200 Tamils, including some women, had been killed.

Another Error of Judgement

In September 1995 the government launched another offensive – and made another error of judgement: it imposed censorships on newspapers and on the reporting of military news from midnight on 21 September. This was tantamount to telephoning the LTTE leadership to warn them that an offensive was imminent. Several towns fell to the army but not as many as might have done had a surprise attack been made.

The deputy Defence Minister, Anruddha Ratwatte, declared that the government would win the war in two or three months. It seemed astonishing that the government had not yet learned that the LTTE regarded such statements as challenges and always retaliated with a spectacular outrage. On this occasion, on 20 October 1995, they blew up two gigantic oil storage depots in Colombo. In the fight that followed with guards at the depots another 20 people died. The loss to the Sri Lankan economy ran into millions of dollars. Simultaneously with this tremendous act of sabotage, groups of Tiger death squads hacked to death ordinary villagers, choosing at random one settlement every day for a period of five days.

Every November, the monsoon reduces conflict and sometimes makes military operations impossible but in 1995 the army ignored the rains and pushed on towards Jaffna and 500,000 civilians – about 60 per cent of the Jaffna Peninsula's population – fled their homes. The capture of Jaffna was achieved but as a result any hope that the LTTE would return to the negotiating table was at an end. During the offensive 466 soldiers and 1,700 Tamil fighters were killed and only 3,000 residents remained. About 400,000 had fled to parts of the peninsula still controlled by the LTTE. The Tigers established a new HQ in the main northern town of Kilinochchi.

Yet again, there was a price to pay for the army's success in their 50-day offensive. A suicide bomber drove a vehicle into central Colombo on 1 February 1996 and exploded it. More than 100 people died and 1,000 were wounded. The outrage ranks high on any international list of casualties from a single terrorist bomb.

After the capture of Jaffna, Mrs Kumaratunga and her ministers did something immensely symbolic. In the Presidential Secretariat in Colombo, Defence Minister Ratwatte presented her with a scroll rolled up inside a red velvet container and dated 'full moon day of the month of Uduwao in the year 2939 of the Buddhist era'. The most significant sentence in the screed was this: 'Your Excellency's rule and authority in Yapa Patuna has been firmly re-established'.

Yapa Patuna is the term employed by conquerors in medieval times. The use of this Buddhist iconography indicated that Mrs. Kumaratunga had conquered Tamil lands and defeated her enemies as Sinhalese rulers had done for centuries. She insisted that the ceremony itself and the language used had nothing to do with conquering, but this was not how the Tamils perceived it. They were angry and offended. The President repeated her plan to transform Sri Lanka into a federation of eight regions, including a Tamil-dominated area, each with wide autonomous powers. This, however, would require parliamentary approval and strong groups of MPs opposed the plan as 'diversive'.

As I commented at the end of my analysis of the Sri Lankan War in *War Annual 7*, 'There is no defence against terrorist driven by such a level of fanatical rage and hunger for revenge'.

The War in 1996–97

In April 1996, the government and the army chiefs did something innovative and intelligent. In Operation Sunshine Two, they cleared a safe passage for civilians in the Tamil-controlled areas, allowing them to return to Jaffna. Openly defying the Tigers, about 250,000 out of an estimated 500,000 people of the Walikamam sector took advantage of the opportunity.[2]

In some places Tamil fighters watched helplessly as crowds of people, carrying their few possessions, flocked along the roads. Short of massacres of their own people on an enormous scale, they could not have stopped the exodus. As they reached safety, the civilians were handed food and various supplies, including basic agricultural implements and seeds.

The chief military spokesman, Sarath Munasinghe, said, 'This is probably the biggest ever setback to the Tigers. It is a turning point. Tamils turning their backs on the Tigers must be more stinging than any military onslaught.'

But the Tigers had not surrendered. They made many hit-and-run attacks and killed numerous members of the security forces to avenge their defeat.

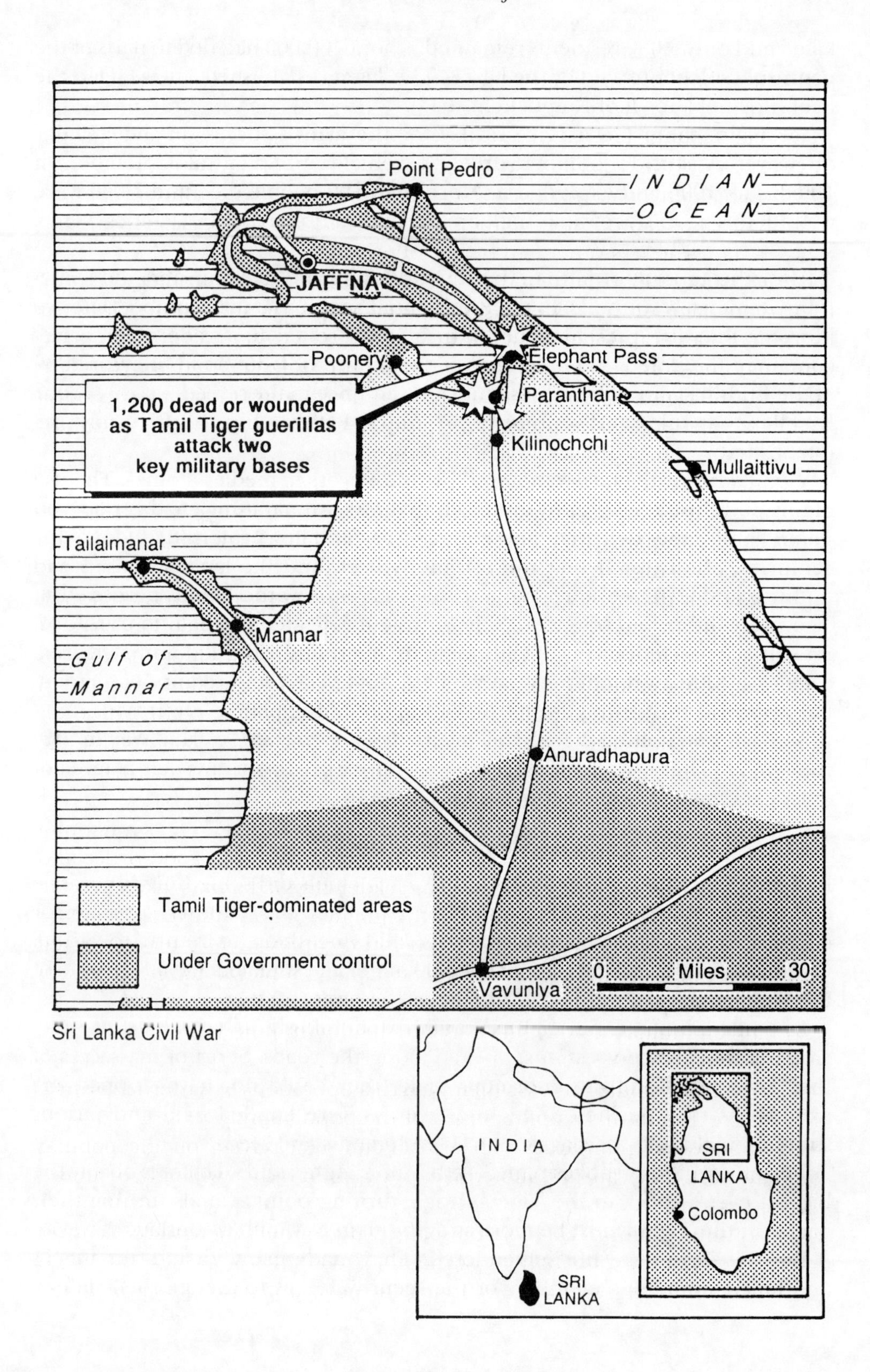

Sri Lanka Civil War

They had already lost the southern peninsula port of Kilali and on 16 May they were pushed out of Port Point Pedro on the northern coast. The fleeing rebels were pursued by helicopter gunships while navy gunboats cut off their escape. Point Pedro was the last important strategic LTTE stronghold on the peninsula and the base of the Sea Tigers, who made a speciality of ramming navy vessels with bomb-laden boats. Of even greater importance, Valvettiturai, close to Point Pedro, was captured; this is the birthplace of the LTTE's leader, Vilupillai Prabhakaran, and it is venerated by his followers.

During May 1996, General Gerry de Silva, who had been leading the anti-Tamil campaign since 1994, was succeeded in command of the army by Lieutenant General Rohan Daluwatte. What these two senior officers said at their handing-over ceremony was significant. De Silva conceded: 'We have defeated the Tigers because they are hiding, not coming out to face us but a political settlement will be needed to end the war.' Daluwatte said: 'We cannot eliminate the rebels. If we are able to push them into jungle terrain we have brought the war to a lower phase of insurgency.'

According to Daluwatte, the strength of the Tamil Tigers had been halved to 8,000 fighters, though diplomats in Colombo could not verify this and believed the number was higher.

Despite their defeat in the Jaffna Peninsula, which was now obvious enough for everybody to see, the Tamils were still receiving substantial supplies of munitions. Previously, the government had ordered new ships for the Sri Lankan navy and in May 1996 five vessels arrived from China. They consisted of a large patrol boat, three fast attack-gunboats and a landing craft. With their customary aggressiveness, the Sea Tigers were ready to retaliate against the government's increased naval strength and on 11 June their frogmen sank two naval patrol boats and damaged a third at the Karainagar naval base.

In fact, the defeat has in no way meant the end of the LTTE as a fighting force. As might have been expected, the rebels withdrew in good order and retained most of their military assets. The leadership had always taken precautions against changes in fortune and in the jungles of Mullaittivu, they possess what is known as One-Four Base Area. It is a complex of 14 inter-connected camps south-west of Mullaittivu and the Sri Lankan forces would need a great and expensive operation to capture it.

Sri Lankan military intelligence was anticipating in June 1996 that:

- The LTTE would increase their guerrilla operations in Eastern Province, where Tamils make up one third of the population.
- They would reinfiltrate the Jaffna Peninsula, relying on the popular support they can undoubtedly expect, given either willingly or under coercion.
- Continued pinprick attacks on security personnel. There were plenty of targets – a full 25 battalions of them.
- Attempt to cut supply lines, forcing the government to use ships (to the

port of Kankesanturai) or aircraft (to Palaly airbase).

The Sri Lankan armed forces' main problem from mid-1996 onwards was shortage of manpower. A leaked report to President Kumaratunga indicated that the defence chiefs wanted the army brought up to its full establishment of 105,000, double the actual figure in June 1996. Diplomats in Colombo said that this large force – which is twice the size of the British army – would be necessary to hold Jaffna and to defeat LTTE in Eastern Province but they doubted that Sri Lanka could afford the great cost, particularly as foreign governments were already reluctant to contribute further large sums of money.

One idea was to re-employ army veterans, even some up to the age of 60, as garrison troops while retraining ordinary infantry as commandos and paratroops. In July 1996 more that 3,000 army deserters responded to an offer of amnesty and returned to their units without penalty, though 7,000 remained at large.

The Army's Greatest Defeat

During the last week of July 1996 the LTTE showed in the bloodiest possible fashion that it was still a powerful fighting force. The rebels inflicted the greatest single defeat of the war on the army by killing 1,200 soldiers in the north-eastern town of Mullaittivu. The soldiers regained control of their base but the army was shocked, as was the entire nation. In retaliation, terrorists planted a bomb in a train in Colombo, killing 70 people and wounding a further 500. The Tamil group People's Liberation Organisation of Tamils (PLOT) was confident enough to reappear now that the LTTE appeared to be weakened. PLOT's leader, Dharmalingam Siddhathen, in a statement to a newspaper, said: 'We stand where we stood 13 years ago. More destruction, more people killed, but we have not moved very far towards finding a solution to the ethnic problem.'

On 26 July, in an operation that was partly inspired by vengeance for Mullaittivu, the army moved south from its key Elephant Pass base to seize Kilinochchi. This military movement sent scores of thousands of civilians fleeing for refuge in churches and schools. By capturing and holding Kilinochchi, the army would gain a tremendous tactical and logistical advantage and would no longer have to make detours by sea and air to supply the northern garrisons. The capture of the strategic town would also enable the army to ferry home across the Jaffna lagoon at least 100,000 refugees from the peninsula.

Much to the High Command's surprise, the Tigers' resistance was so fierce and their fixed defences were so strong that the army's advance had stalled by mid-August, three miles short of the town. The army's artillery fire was heavy but in their deep bunkers the Tamils suffered few casualties. Meanwhile, because the fighting spread across all routes from further south, humanitarian supplies could not get through from Vavuniya to the hundreds of thousands of refugees. Both sides suffered considerable casualties.

From merely holding the army, the Tamils went on the attack towards the end of September 1996. The military said that the attack was 'repulsed very effectively with heavy casualties among the terrorists'. The stated figure was 325 killed. Perhaps more significantly, a barrage of mortar shells landed among an infantry unit, killing 97 and wounding 250 others.[3]

Despite the Tigers' resistance, the army recaptured Kilinochchi on 29 September. They had launched a new offensive from Paranthan, encircled and then crushed the defences. It was the eighth day of 'Operation Sathjaya' Phase III. About 20,000 troops took part in the final attack. Given such a weight of manpower, artillery fire and rockets fired from gunships, the final result was inevitable. The army's official casualty figures were 269 men killed and 311 wounded, with 700 rebels killed.

Before dawn on 10 January 1997, the Tigers were back in action, making massive raids and artillery attacks against the Paranthan and Elephant Pass army bases. Again, both sides claimed success and it was certainly true that the LTTE forces withdrew after the raids and that they suffered losses, though probably not the 350 killed and 700 wounded that the defence ministry claimed. Most of the military casualties resulted from 122mm guns fired by the Tigers. LTTE demolition squads caused great damage, wrecking at least 15 of the army's heavy guns and smashing a 6-mile bunker line. The Tiger offensive was the most destructive since the rebels had wiped out the garrison of 1,200 troops at Mullaittivu. If there was a lesson to be learned from the chaos caused by the determined raids it was that the troops were stretched too thinly in hostile territory and so vulnerable – but that lesson should have been learned many years earlier. The army conceded, in effect, that its lines were thinly held when it reported that a major search was under way to look for wounded and small groups of troops who might have run away and then been cut off. The army's morale, which had recovered after the capture of Kilinochchi, was badly hit by the Tamils' twin-attacks on Elephant Pass and Paranthan.

Sri Lanka's Special Boat Squadron (SBS)

The visit to Sri Lanka on 18 August 1996 of the US Co-ordinator for Counter-Terrorism, Ambassador Philip Wilcox, together with a team from the State Department and Pentagon, was supposed to be a secret. When it became known that Wilcox and his team were having high-level talks with Sri Lanka's defence and foreign affairs departments it was not long before the specific reasons for the Americans' visit became clear.

The Sri Lankan army is deeply critical of the Sri Lankan navy, accusing it of failing to provide adequate support, especially during the Mullaittivu disaster. Apparently the navy, engaged in evacuating wounded soldiers from a beach came under fire and abandoned the soldiers. Also, according to army officers, the LTTE's Sea Tiger raiders are much more daring that the Sri Lankan navy's special units. They conceded that the navy could not introduce the equivalent of the LTTE's Sea Tiger suicide squads but insist that it should be able to counter the activities of the LTTE's underwater demolition units.

Early in 1996 the navy set up its Special Boat Squadron (SBS), which it modelled on the British Royal Marines units of that name. One purpose of the Wilcox visit was to see if the US navy's SEALs could also provide a model. The conception of SBS squadrons was long overdue, for the Tigers had long dominated the coastal areas of northern Sri Lanka. Wilcox and his officials also discussed the private involvement of the US company, Military and Professional Resources. This firm had proposed, perhaps on government advice, courses of training for élite Sri Lankan units, including the SBS. They could do what the US government could not do – but would like to see done. The Americans, while understanding the Tamils' desire for an independent state, regard many of its activities as outright terrorism. Bombs in Colombo could hardly be seen as anything else.

Officially, the US government denies involvement in the civil war but it does have links with the Sri Lankan government in a programme known as Joint Combined Exchange Training.

In the end, the army will have much to say about the SBS's training and performance since the unit's commander is an army brigadier, appointed personally by President Kumaratunga.

Sri Lankan Intelligence

During 1996 the government appeared to hand the Tamil opposition a considerable intelligence advantage when it forced through a complete reshaping of the National Intelligence Bureau (NIB). The purpose was understandable enough – to 'eliminate weaknesses' – but diplomats in Colombo said that the reforms, even if they were necessary, were shortsighted and too sweeping.

The NIB is an extremely busy agency, as might be expected of a nation at war for so long. While its main duty is to spy on the Tamil rebel organisations inside Sri Lanka, it must also closely watch the LTTE's arms procurement operations and source of funds. Because of this the NIB has agents in Britain, India, Australia and the United States. The London office is the most important after Colombo.

Clearly, the NIB has had many failures or it would have foreseen some of the terrorist attacks. Its assessments of LTTE capabilities have been inadequate, even at times absurd. Because of all this the People's Alliance government has said that some of the intelligence chiefs are 'politically unreliable', a euphemistic way of accusing them of being traitors. It would be surprising if some agents had not been coerced, bribed or blackmailed into helping the LTTE, especially through threats made against families.

During 1996 the government moved many senior NIB officials to new posts while some were retired. Undoubtedly, some changes had been politically motivated; for instance, some officers appointed by the United National Party who were UNP members were removed to make way for People's Alliance nominees. As a French diplomat told me, 'This is no way to run an Intelligence service during a war'.

LTTE Official Casualties

In mid-January 1997 LTTE, through its offices in Madras and London, issued statistics for fatalities suffered in the period November 1982 to November 1996. I have asked Sri Lankan Red Cross officials, two foreign diplomats in Colombo and a Sri Lankan military journalist for their opinion about the accuracy of the figures and all said that they were credible. The total number of dead was 9,301, including 1,079 women fighters. The majority died in raids, patrols, ambushes and 'stay-behind' suicide missions. Of the total, 2,727 were killed in major operations, or what the LTTE would regard as battles, offensives or large-scale raids.

The largest single loss was at Elephant Pass, in 1991, when 602 Tigers were killed. An over-ambitious attack on the army base at Pooneryn, Jaffna Lagoon, in November 1993 resulted in 459 deaths. When the LTTE demolished the army base at Mullaittivu, in July 1996, and killed 1,200 government troops, 314 Tigers were lost. Despite the heavy toll in Tigers' lives, the LTTE considered Mullaittivu a great triumph. The defence of Kilinochchi, in July–September 1996, cost the LTTE 241 fighters.

While the Sri Lankan defence ministry has issued many casualty lists it has never produced a statistical analysis but the same sources which considered the LTTE's published statistics produced estimates which give an average of 18,000 deaths for the security forces.

The interesting question is: why did the LTTE release its statistics at all? The answer must be that it wanted to show the Hindu Sri Lankan public that its losses were calamitous and, in proportion to the Tamils' losses, they should be regarded as intolerable. The death toll of more than 9,000 Tamils also indicates the intention of the LTTE to fight until they are given an independent homeland. The more men and women fighters they sacrifice, the greater their determination to achieve their aim. In short, the publication of what must be considered reliable statistics was a propaganda exercise.

It will be noted that the Sri Lankan army rarely gives figures for the number of Tamil wounded and taken prisoner in action. This is because the army rarely takes prisoners, wounded or otherwise. To a man, the Sri Lankan security forces regard the Tamil fighters as vermin to be destroyed. In their public utterances, senior officers and military spokesmen refer to them in this way. Apart from army ruthlessness, Tamils are intent on not being taken alive and they carry poison pellets to ensure that this does not happen.

1997 – The Year of Hopelessness

Throughout 1997 both sides of the dispute made advances into 'enemy' territory – and both sides withdrew. Both suffered more casualties. The government offered 'concessions' and asked for talks. The Tamils sometimes accepted the idea of talks – and then did not turn up. They could not always be blamed; on occasions the government was not sincere but merely trying to gain international credibility.

In August 1997 a diplomat with much experience in Sri Lanka told me: 'This is a country without trust. It has a stab-in-the-back culture. It is a country where so much blood has been spilled that each side demands that more be spilled in reprisal. No matter what deal is eventually worked out, and there will be a deal, neither the Tamils nor the Hindus will accept it for long. Perhaps the best solution is for the United Nations to impose partition and send in a force of Blue Helmets. But they would be partitions's first victims. In 300 years' time Sri Lanka will be as Northern Ireland is today, without trust, without tolerance.'

This is a sad conjecture. Having spoken at length with Tamils and Hindus, including their politicians and military leaders, I must agree with the diplomat.

References

1. Mrs Kumaratunga is the daughter of the former Sri Lankan Prime Minister, Mrs. Bandarnaike, who, in 1948 helped to make Sri Lanka, then Ceylon, independent from Britain.
2. 'It was the first smart move by any government since the Tamil-Buddhist confrontation began', a Western diplomat told me.
3. According to a Red Cross report on 25 April 1996.

Earlier editions of *War Annual* dealt in detail with the following aspects of the civil war:
The role of Tamil women as Tiger fighters, the 'Freedom Birds'.
The Indian army's abortive attempt to restore peace in Sri Lanka.
The Maoist terrorist group, the *Janata Vimukti Paramuna* (JVP) or People's Liberation Front and the government's successful campaign to crush it.
The first Battle of Elephant Pass.
The training of Tamil fighters.
Operation Leap Forward and Operation Tiger Leap.

27

Sudan Civil War

A TURNING POINT

In 1955, just before Sudan gained its independence from British-Egyptian rule, the garrison of the town of Torit, in southern Sudan, mutinied when ordered to march to Khartoum for a new posting, leaving their arms behind. The southern troops, mostly Christian, believed that this was a ruse to capture and massacre them and they mutinied. We now know that their fear was justified.

The new military government declared war on the south and conflict continued sporadically for years. A military coup in 1971 brought to power General Gaafar Nimeiri, whose one major success was in negotiating a peace with the southern leaders in 1972. However, the armed forces and extremist Muslims in Khartoum continued to harass the Christians and animists of the south until, in 1983, they formed a self-defence force, the Sudan People's Liberation Army (SPLA), also known as Anganya meaning 'venom of the viper'. Its leader then, as now, was Colonel John Garang. Nimeiri had promised Saudi Arabia that he would impose *Sharia* law on Sudan, with all its extremes of punishment. Within a short time floggings and executions became commonplace. In 1985 another general, Swar el-Dhahab, seized power and held it until 1988 when he, too, was deposed by a paratroop brigadier, Oman Hassan al-Bashir. A Muslim extremist, Bashir ruled through a 15-man military junta which greatly increased military pressure against the SPLA. Until 1989 the world may not have known just how many Sudanese had died in the civil war but in that year the UN put the figure at 500,000.

Several international statesmen made attempts in 1993 to persuade Bashir to adopt moderate policies but all were rebuffed. In January 1994 Bashir, determined to crush the SPLA and Garang, shipped thousands of men with armour and artillery down the Nile to a base close to Juba. Garang fell back but was not defeated. The Muslim troops killed many thousands of civilians but few SPLA troops.

Sudanese mullahs were inflaming unrest in Egypt and Bashir established three camps to train Egyptian Islamic fanatics in terrorism. Sudan was also exporting arms to militants in Algeria. In 1994 Eritrea broke off diplomatic relations, alleging that Sudan was training Islamists to destabilise it. Uganda expelled Sudanese military observers while Ethiopia strengthened its forces on the Sudan border.

In March 1995 the UN Human Rights Commission condemned the Sudanese government for abuses, including summary executions, slavery and systematic torture. The Commission reported that Christian children were being sent forcibly to remote camps for indoctrination and forced conversion to Islam. Amnesty International reported that thousands of southern Sudanese had been sold into slavery while African Rights accused the government of an 'horrific range of abuses'.

There was no doubt that Bashir and his military junta, unable to defeat the SPLA in the field, were trying to destroy southern society. They had certainly killed many of the people – 1.3 million by the end of 1996, according to UN estimates.

The War in 1996–97

In May 1996 Baroness Cox, the president of Christian Solidarity International, brought to the world's attention the appalling situation of the oppressed people of southern Sudan, an area she knows personally. Part of her report reads:

> The government's policy towards the people of the South and the Nuba mountains is tantamount to genocide, by means of terror, war, slavery, the mass displacement of population and the manipulation of aid. In particular, widespread, systematic slavery continues on a large scale in government-controlled areas of Sudan. The raids by government troops and government-backed militia against African towns and villages of the South and Nuba mountains are accompanied by atrocities, torture, rape, looting and destruction of buildings and property. Those not taken into slavery are generally tortured and killed.[1]

This graphic description points to a particularly vicious form of warfare, defeating an army by attacking the families and communities whom the army was protecting.

In September 1996 Jemera Rose of the Human Rights Watch organisation visited the garrison town of Juba. She was barred from going into the market and from staying at the UN compound, as she had planned; instead she was forced to stay at a government-controlled hotel. Ms Rose was not allowed to speak privately to anybody, even the local Roman Catholic Bishop. She was constantly accompanied by security personnel and when she protested they escorted her to the airport for deportation.[2]

In Khartoum coup attempts took place during 1996, the first in March, when many army officers were arrested. In another coup in August a further 65 officers were arrested. On 26 October, Egypt's official Middle East News Agency reported that a group of ten colonels had mounted a coup; a later report said that these men had all been executed. In yet another development, on 22 October a Sudanese air force pilot defected to Saudi Arabia in his Chinese-built F-6 fighter and sought political asylum. His choice of sanctuary was puzzling, since Saudi Arabia is one of the few countries with any sympathy for the Bashir regime.

A few days after the pilot's defection, the former prime minister of Sudan, Sadiq al-Mahdi, overthrown by Bashir in 1989, broke a long silence. Writing in an Arabic newspaper, he said that a 'military struggle may be inevitable if the fundamentalist regime is to be overthrown'. While al-Mahdi's government was an elected one he, too, was a fundamentalist. His comments were interpreted as encouragement for officers of Sudan's armed forces to rise up against Bashir.

The Rebel Offensive

Various opposition groups, both political and military, came together under the umbrella National Democratic Alliance (NDA) in order to bring about what its leaders called a popular uprising and to open up as many fighting fronts as possible. Eritrea and Ethiopia gave support to the NDA. Both countries had long provided refuge to opposition groups. The new rebel offensive began on 28 December with an attack on a mechanised brigade near the Eritrean border.

On 12 January 1997 the campaign intensified, in two sectors. One was near the south-eastern border with Ethiopia, the other 300 miles away to the north-east, near Kassala, on the border with Eritrea. Bashir declared a general mobilisation for the 90,000-man army and the paramilitary Popular Defence forces of 15,000 and its reserve of 60,000. Sudanese universities closed as students and staff received call-up papers for military service. Bashir took the war threat seriously, for he announced on national radio, 'the decisive battle has begun'. He personally bade farewell to a long convoy of troops and supplies leaving Khartoum for the east.

At the same time he reproached the Ethiopian leadership for their 'ingratitude'. He was referring to Sudan's help for the Tigrayans who had fought a war of liberation against the Marxist regime of President Mengistu, which ended in 1991. The Tigrayans later became the ruling ethnic group in Ethiopia. Bashir said:

> When you arrived starving, we gave you food. When you came to threaten us, we calmed you. Now that you come as invaders we are going to give you a lesson you will never forget.

Ethiopia and Eritrea officially denied involvement with the Garang's SPLA and the NDA but all African diplomats in those two countries and in Uganda and Kenya know that army officers and NCO instructors from Ethiopia and Eritrea work closely with the rebel troops. Some actually lead Sudanese rebel units.

Eritrea gives the NDA a home and supplies it with weapons. The United

States, which has long denounced Sudan as one of the major terrorist nations, has made no secret of the fact that it supplies 'defensive' weapons to Eritrea, Ethiopia and Uganda to help protect them from Sudan. These arms reach the rebels if not with the Americans' connivance then certainly with their knowledge.

Sudan's parliamentary speaker, Hassan al-Turabi, conceded on 15 January that the army garrisons along the eastern border with Ethiopia had been taken by surprise. He said that the army had only four battalions deployed along the frontier, mainly to combat smugglers. At this point, the Bashir regime was in near-panic and calling on fellow Arab governments to support it. Bashir played on their paranoia that Garang's SPLA was a pawn of Western powers trying to break up Sudan. A few, including Libya, Syria and Jordan, supported Sudan with strong statements but none offered aid.

The SPLA and its allies claimed to have killed 1,260 Sudanese troops as they moved steadily towards the strategic town of Damazin, capital of Blue Nile State. Damazin is the site of the great power station that supplies Khartoum with electricity and water. Of course, this claim was denied by the Sudanese information minister, al-Tayeb Muhammad Khair, who said that now more than ten soldiers had died. Whatever the true figure, the 'battle for Damazin' went on for 12 days, and between 12–24 January the rebels occupied 1,200 square miles or 15 per cent of Blue Nile State. The towns of Kurmuk and Qeissan were particular prizes.

According to the Sudanese information ministry, the army had killed 300 Ethiopian troops and he said later they were specifically identified as Tigrayans; the Tigrayans are the ruling ethnic group in Ethiopia.

Strategically, the rebels needed to cut off Khartoum's electricity supply, capture the Nile dam at Roseires or Port Sudan to Khartoum. Any of these achievements could bring about a successful coup against Bashir. But while this outcome could be predicted the longer view is clouded. As *The Economist*, London, has said: 'The current north-south alliance would be unlikely to survive. Sudan, huge, disparate and surrounded by interfering neighbours would find it hard to stay together, democratic or peaceful.'[3]

References

1. Bernard Levin, a columnist for *The Times*, London, used much of Baroness Cox's information in an article headed 'A Slave State of our Time: Southern Sudan is being laid waste by the Khartoum Government – and all in the name of a merciful God'. 31 May 1996). Levin wrote, 'Food and drink are available but they are deliberately denied. Hunger is used as a weapon and thousands of victims of the government's genocidal plan rot and die.'
2. Human Rights Watch produced a 340-page report about conditions in Sudan. It stressed the widespread use of torture and murder as a means of oppression in the south and gave many examples of Christians being captured by government troops and sold as household slaves in the Muslim north.
3. On 20 January 1997, *The Economist's* analysis of Sudan made one of the first mentions of the Beja Congress Armed Forces, a new part of the National Democratic Alliance.

28

Child Soldiers: Cheap, Obedient – and Hungry

In 1995 the UN Secretary General Boutros Boutros-Ghali asked the distinguished Gracha Machel, wife of a former president of Mozambique, to investigate the systematic use of child fighters around the world. Mrs Machel spent a year or more visiting Angola, Cambodia, Rwanda, Burundi, Liberia, Sierra Leone, Sri Lanka, Lebanon, Burma (now Myanmar) and Northern Ireland. Difficulties were put in her way but through sheer persistence and with considerable courage Mrs Machel submitted her 94-page report in November 1996. She called it *The Impact of Armed Conflict on Children.*

In brief, she blamed the international arms trade for the easy availability of cheap assault rifles, together with the cruel exploitation of easily-influenced children by many warlords. She pointed out that the mass availability of weapons could transform any local conflict that might once have been settled with relative ease into 'a bloody slaughter'. Weapons are very cheap indeed. In Uganda, Mrs Machel said, an AK-47 could be bought for the cost of a chicken and in northern Kenya for the price of a goat.

In places the Machel report stated the obvious. For instance, 'Previously, the more dangerous weapons were either heavy or complex, but these guns are so light that children can use them and they can be stripped and reassembled by a child of 10'. The Soviet AK-47 was the preferred weapon in civil wars and ethnic conflicts, Mrs Machel reported, and was widely used by guerrilla movements. There was some irony in this observation, though Mrs Machel ignored it: Mozambique's national sports flag bears an AK-47 as its symbol.

Quoting research by the UN Children's Fund (UNICEF) Mrs Machel stated that between 1975 and 1995 two million children were killed in armed conflicts, with another six million seriously injured or permanently disabled. Land mines caused many of these casualties.

Mrs Machel encountered many military commanders who told her that they preferred to recruit child soldiers and she was in no doubt about their reasons. The children were 'more obedient, do not question orders and are easier to manipulate than adult soldiers'. But they are not necessarily fighting all the time. Many are employed as messengers, cooks, porters and increasingly as spies.

In the spirit of a serious researcher, Mrs Machel produced 24 case studies of child soldiers aged between 8 and 17 years of age. She estimates that they make up 34 per cent of Afghanistan's armies, partly because in a land virtually without schools there is nothing else for the boys to do but join an army. However, girls are recruited in several countries. Two of the studies concern, respectively, Lebanon and Sri Lanka, where terrorists recruit and then indoctrinate children aged between 12 and 15 for suicide bombings. Girls are preferred for these missions as they are less likely to be suspected by the security forces.

In Burma, Mrs Machel found that adolescents aged between 15 and 17 were forcibly conscripted from their schools, class by class. In Northern Ireland, she suggested, children were incited to antagonise peacekeeping troops by throwing stones, bottles and petrol bombs.

In concluding her report, Mrs Machel urged that a worldwide campaign be organised to prevent the recruitment of children under 18 into the armed forces. This could be achieved in part if nations would adhere to the Draft Optional Protocol to the Convention on the Rights of the Child. The report leaves readers with the feeling that Mrs Machel does not expect rapid action and neither do I. Children are in plentiful supply, they are generally hungry and they will do anything for food. They will even slaughter people known to them.